Chaos and Intoxication

Why are people often so unpredictable? Why do they do things which can cause great personal harm even when they know this to be the case? *Chaos and Intoxication* examines the nature of drug and alcohol use, seeking to address these and many other enduring questions through a detailed discussion of the chaotic nature of human existence.

Chaos and Intoxication explores three general areas, the first of which is neurobiology and genetics. The evolution of the mind is examined from a Darwinian perspective, drawing attention to the way chance and uncertainty in development are structured by natural selection. Key findings from current biological and medical research are reviewed, the interrelationship between genetics and experience is explored, and Gerald Edelman's theory of the evolution of the mind through natural selection is discussed. The second theme, cognition and collective action, is considered in the light of evidence indicating that the way we think is also subject to natural selection. Furthermore, it is argued that there is a meaningful distinction between reason (adaptive rationality) and formal rationality. Finally, recent research into chaos theory, order and complexity is reviewed and the preceding discussions of biology, psychology, the role of society and collective action are brought together into a general theory of chaos and human nature.

Alan Dean is Lecturer in the School of Social and Political Sciences at the University of Hull.

Chaos and Intoxication

Complexity and adaptation in the structure of human nature

Alan Dean

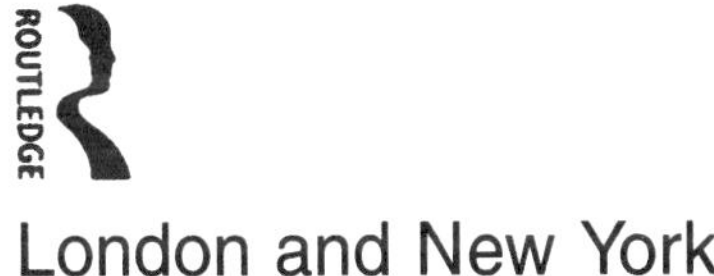

London and New York

First published 1997
by Routledge
11 New Fetter Lane, London EC4P 4EE

Simultaneously published in the USA and Canada
by Routledge
29 West 35th Street, New York, NY 10001

Typeset in Times by
BC Typesetting, Bristol
Printed and bound in Great Britain by
Hartnolls Ltd., Bodmin, Cornwall

British Library Cataloguing in Publication Data
A catalogue record for this book is available from the British Library

Library of Congress Cataloging in Publication Data
Dean, Alan, 1950–
    Chaos and intoxication: complexity and adaptation in the
structure of human nature/Alan Dean.
        p.   cm.
    Includes bibliographical references and index.
    1. Alcoholism.   2. Human behavior.   3. Chaotic behavior in
systems–Social aspects.   4. Chaotic behavior in systems–
Psychological aspects.   I. Title.
HV5035.D4   1997
362.292–dc21   97-2203
                                                               CIP

ISBN 0–415–14614–3 (hbk)
ISBN 0–415–14615–1 (pbk)

This book is dedicated to the memory of my brother
Tom Dean

Contents

Acknowledgement

With many thanks to Mitzi Wakefield without whose endless support and encouragement this book would have taken far, far longer to write.

Towards a chaos theory of human nature

Over many years of conducting research I have often wondered what it is that gives rise to both the vast diversity and the universal similarities which exist between individual people, societies and cultures. For those of us who have been lucky enough to travel far abroad and experience different cultures, there will almost certainly have been moments when everything encountered seemed alien and beyond comprehension, whilst at other times the familiarity felt when amongst culturally very different people may have been quite striking. And distance from home need not have played a part in providing these insights. At times I have felt more at home in the company of indigenous peoples in various mountain regions of the world than I have amongst some groups of people from my own country. Even the most cursory observation of different people throughout the world brings one to the irresistible conclusion that, although as a species humankind has many different ways of organising itself in terms of detail, there are many aspects of human life which are universal in nature. Wherever you go people are engaged, for example, in intimate relationships, organise themselves into kinship groups which inform the ownership of property, the rules of inheritance and the care of children, and share an aptitude for engaging in conflict if necessary to protect their territory or property. I am sure there are many other aspects of human life which the reader could list which have to some marked degree a universal character. Of course I am not saying that each of the elements listed above will be identical in form, but rather suggesting that they represent a list of interests which will be encountered in one form or another universally. That is, with respect to these and many other aspects of human action there will exist both universal and individual elements. So, notwithstanding any minor disagreement which may stem from my selection, the similarities and equally obvious differences (such as dissimilar forms of marriage) which arise between different cultures mean that the question of how these dissimilarities come about needs to be asked. What is it about the deep structure of human existence which allows for diversity and yet often reproduces great similarity?

When I first began to address these issues it was with respect to questions of morality. Why, I wondered, is it, say, immoral to take a human life in consequence of an extra-marital affair in our society whereas in some others this is held to be morally correct behaviour? Why have some societies engaged in ritual cannibalism, when to those of us in Western Europe this is an abominable practice? There is of course a trivial answer to these questions in that one could merely adopt some relativist position and argue, philosophically, that what is good or bad is not fixed, but is determined within specific cultures. I have always thought that such reasoning is only removing the problem to a different level, because at some point those holding the relativist view will be required to reveal the nature of the mechanism whereby observable outcomes arise. That is, with respect to human nature, you just cannot say that events happen, you need to demonstrate how they happen. To explain why both diversity and similarities arise within groups of biologically closely related people what is needed is a theory of human nature which accounts for both the universality which we observe and the differences. We need a general theory of human nature which accounts for all the observed data, and not merely some convenient sub-set. What we need to know is what is it that leads to predictability (similarities) and uncertainty (dissimilarities) with regard to behavioural, social and cultural outcomes?

I wished to address these issues directly for a long period of time, but before I could do so I needed a topic upon which a suitable analysis could be based. In this regard the study of drug and alcohol use seemed an entirely appropriate starting point, because, within this area, there seemed to be a considerable amount of moral confusion and uncertainty. On the one hand it was apparently generally okay within British society to become extremely drunk on alcohol, but generally not okay to use large quantities of cannabis, cocaine or heroin. In fact, the reality is even more complex than this as drunkenness is not always acceptable, but is instead socially a closely monitored and defined condition. Although the rules surrounding permissible intoxication are extremely complex, it is mostly true to say that being drunk through alcohol use is subject to less social censure when it arises within an accepted collective celebration, such as a Christmas party or a bachelors' night, than as a result of more casually based drinking during, say, a Saturday lunchtime. And, to extend the area of interest to include questions relating to cross-cultural issues, it may also be asked as a research question why it transpires that it is morally acceptable to consume coca leaves in certain South American countries, when it is not so in the British Isles?

In reflecting upon these issues I realised that in seeking the answers to questions concerning the deep structure of human nature, I appeared to be directed towards mundane affairs rather than great issues of the day. However, in the subjectivist tradition, I felt that the study of the mundane in our lives was legitimate as this approach will often reveal the greatest insights. Indeed, these interesting dissimilarities in practice and social responses surrounding drug and alcohol use seemed to provide an excellent means to explore questions relating to personal, social and cultural diversity and universality.

Reflecting on these matters it seemed that this would be a relatively easier topic to pursue in practice through research if the related empirical work of such a study were based in a community which was to a significant degree culturally distinct from any nearby metropolitan area. It seemed evident that seeking out the cultural and historical basis of patterns of drug and alcohol use (an approach required if findings were to be grounded within the interests of a specific community) would be more achievable the more homogeneous the selected community was in terms of social and cultural practices. Attempting such a programme of research in, say, London, a centre of enormous cultural diversity, would be more difficult (if not impossible) than undertaking the same task in a small rural community.

Through this process of reasoning I eventually began a three-year period of research on alcohol and soft drug use in the Western Isles of Scotland. During this period, which began in the late 1980s, the alcohol and drug-using practices of young people were examined within the context of both contemporary and historical concerns. Once completed, the research illustrated that although continuing practices were informed by contemporary concerns, they were also in important respects grounded in a tradition of intoxication within the parent community (Dean 1990, 1995a, 1995b, in press). It seemed clear that, notwithstanding any claims for a biological basis to alcohol and illicit drug use and misuse, these practices were most certainly formed within the context of social and cultural concerns.

This might have been an end to this series of studies except that finishing the research led to a move to East Yorkshire. Through fortuitous contacts I encountered highly unusual forms of drug use involving veterinary products amongst young people in local rural communities. This was an unexpected phenomenon and quite remarkable in many respects. Why should young people wishing to experiment with drugs engage in such exceptional forms of use? In seeking to unravel these events in terms of their antecedents there grew an analysis which seemed to point, at least in part, to the importance of social and geographic space in determining outcomes. There thus developed an idea which seemed to suggest that what people do arises from the resolution of many different influences upon their lives. What happens to someone can depend upon where they live, who they socialise with, the history of intoxication within the resident community, and many other elements which comprise the world we live in (see Dean 1995b). Of course the unique nature of individual subjects themselves could not be bracketed out of such an analysis, because factors such as personality or willingness to take or seek out risks is also important. In the end there arose an analysis which described a multi-dimensional array of influences within which the conscious (and pre-conscious) intentional self existed and from which actions emerged. The next task was to make some sort of sense out of this complexity. It certainly was not acceptable to present a description of this hugely complex universe of experience and events as a finished product.

A clue to resolving the problem of adequately describing this multi-dimensional context of intoxication presented itself directly from the very complexity which had emerged. It seemed a good starting point to assume, as a working hypothesis, that the underlying structure of such a complex system should be the same as that existing for any other such complex system. And in thinking about this it seemed evident that if it is possible to describe weather systems and other unpredictable dynamic systems by way of an emerging science of chaos, then perhaps there lay the answer: intoxication occurs within a chaotic system. The fact that patterns of intoxication are unpredictable in the way weather systems are, also strengthened my belief that this might be a productive line of thought.

Of course the implications of engaging in the task of presenting a theory of intoxication from the perspective of chaos theory meant that it needed to include every dimension of such use, and not just focus on, say, biology or psychology. Chaos theory describes deep underlying structures and not surface impressions. In consequence I would need to start with the deepest aspects of use to be found at the level of neurophysiological structure, and end with phenomenon at the social and cultural levels – no easy task, but, I felt, if a beginning is ever going to made with respect to building a unified theory of intoxication by way of examining a broader interest in human nature, then the task will have to be undertaken at some point, so why not now?

This last comment draws out clearly the broader context of this current book. In many ways the use of intoxication as a substantive structure for the book is merely a device. Drug and alcohol use is a good medium within which some fundamental aspects of human nature can be explored. As the subject has attracted much attention from biologists, medical scientists, psychologists, economists, anthropologists and sociologists, there exists perhaps a unique body of knowledge on a specific universal aspect of human nature: the pursuit of intoxication. There can be very few other aspects of human action which have been so comprehensively researched. What this meant was that I could engage in an analysis which incorporated data from across the all the natural and social human sciences. In consequence, although this book is substantively about intoxication, in terms of theoretical development it is focused more broadly upon the nature of human existence.

Hence, as you read through the book, there will be many occasions when it seems that drug and alcohol use does not feature at all. Be patient, because in the end it is all tied together and I hope new insights will have been made available with respect to this important area of human activity. But, notwithstanding this, the broader project is beyond any narrow discussion of intoxication. The completed book, I hope, says something useful about the deep underlying structure of human existence.

Being based across many different scientific disciplines there are parts which may contain difficult material not before encountered by some readers. When this occurs I would urge patience and counsel a little perseverance, particularly for those of you closely involved with drug and alcohol misuse. I believe it will benefit the field enormously if natural scientists gain insights from the social sciences, and vice versa. In this respect I hope this book makes a modest contribution.

The book is split into three general areas. Chapters one to three cover recent findings from neurobiology and genetics. Within these early chapters there is developed a Darwinian perspective on the evolution of the mind which draws attention to the way chance and uncertainty in development are structured by the processes of natural selection to give rise to uncertain outcomes at the level of the individual. More specifically, chapter one reviews key findings from current biological and medical research. Chapter two looks at the interrelationship between genetics and experience and examines the relative contribution of nature and nurture to the processes of human development. Chapter three provides a detailed discussion of Gerald Edelman's theory of the evolution of the mind through natural selection.

The second main component of the book comprises chapters four, five and six, where the analysis developed earlier is pursued with respect to cognition and collective action. Evidence which indicates that the way we think is also subject to natural selection, both historically and within the contemporary world, is discussed in detail. Particular attention is focused on an analysis of the difference between reason, an adaptive rationality formed through natural selection and formal rationality which, it is argued, arises collectively within logical and scientific discourses.

Chapters seven and eight review recent research into chaos theory, order and complexity. The principles of chaos theory are presented in chapter seven, whereas in chapter eight a review of the ways that chaos theory can be used to provide insights into social systems is discussed. In chapter nine the preceding discussions of biology, psychology, the role of society and collective action, and the principles of chaos theory are brought together into a general theory of chaos and human nature, from the perspective of intoxication. The way that order arises within chaotic systems through natural selection is discussed. The final chapter reflects on the possibility that human existence is indeed fractal in structure.

This is a broad project, so in this book the focus is on providing an overview of what a completed chaos theory of human nature may need to incorporate. Others may have a different view, but hopefully this book will be found useful in providing at least one starting point for such an endeavour.

Chapter 1

The embodiment of intoxication

A literature search carried out at the Project Cork drugs and alcohol database at the US Dartmouth College in April 1996 gave rise to 791 references in response to the keyword genetics, 231 to psychology and only 33 to history, culture and society. The 1995 Annual Scientific Meeting of the US College on the Problems of Drug Dependence showed a similar trend towards research of a medical, biological or genetic nature. These observations show that there still exists in the 1990s a dominant emphasis within drug and alcohol research towards the search for genetic markers, gene loci and regions of neurobiological reactiveness within the brain which can explain individual differences in the use of mood-altering substances. In many respects this is perfectly appropriate as it is clearly the case that whatever else human beings may be, they are biological entities and as such any comprehensive explanation of human action, at the individual or collective level, needs to be embodied. That is, as Edelman (1992: 239) has stated, thought, and thus intention, is not transcendent but 'depends critically on the body and the brain'. For Edelman the mind is embodied in that evolution and developmental processes give rise to a particular brain morphology from which particular patterns of thought arise: 'Gestalts, mental images, bodily movements, and the organisation of knowledge must all to some degree be the result of evolutionary and developmental constraints [on the morphology of the brain]' (p. 239).

It is too soon in this discussion to expand these difficult ideas further; Edelman's work will be returned to later. But this early reference to his enormous contribution to our understanding of the relationship between brain, mind and body makes clear the need to consider brain chemistry, morphology and neurophysiology in deliberations on alcohol and drug use. In this regard the preponderance of medical and biological research on drugs and alcohol is well justified. However, it is less satisfying that most research, in attempting to reduce the pursuit of intoxication to objective biological processes, appears to dissociate brain from mind and treat substance problems as merely a hardware problem. On reading much research in the area it appears that there is some real belief that if a particular gene locus or neurobiological pathway which gives rise to, say, cocaine dependence, can be identified then such problems could be overcome through specific medical treatment. I realise that this is in some respects an oversimplification of medical and biological research, but it is not exactly completely inaccurate either. Although reviews of current and recent work such as the excellent paper by Edwards and Gross (1976) draw attention to the complex biological, psychological and social nature of alcohol and drug dependence, most primary medical and biological research tends to bracket out culture and intention in pursuit of a reductionist truth bound to notions of disease.

This is not to say that such research should not be encouraged. The task of embodying human action cannot proceed without such primary investigations. The criticism is rather broader than another call for more ethnography, but is instead the beginning of a critique which will argue that we need more than occasional review papers, as scholarly and important as many are. What is required is a new paradigm which embodies human action such that each level of existence from culture to biochemistry can be understood, conceived of, in terms of the other. I shall attempt something along these lines in this current text as a journey from genes to brains through cognition to consciousness and collective action unfolds. The story will be incomplete as the priority now is to get down on paper a broad sketch of the territory, but I hope it will provide some new ways of looking at a classic problem of human action: why we like getting intoxicated and why some people like it more or less than we do.

BIOLOGY AND IDEAS ABOUT ADDICTION

The most appropriate starting point for this discussion can arguably be what is known about the extent to which alcohol or drug problems are determined by biological predispositions. In doing so, the work of Jellinek (1960) provides perhaps the best introduction. Although not referred to as frequently as it once was, this work still stands as, in many ways, the most complete statement of alcoholism-as-a-disease that has yet appeared. It is complete, not in the sense of detail – there have been many developments in molecular biology which Jellinek could not have anticipated – but with respect to the extent to which what he called *alcoholism* (a term still favoured by those who depict substance problems as a disease), arose from a complex interaction of biological, psychological and environmental elements. No simple biological determinism can be discerned, but instead a wide-ranging analysis in which ideas about culture (when comparing French and Italian wine-drinking customs) stand alongside those concerning the nature of craving.

Jellinek described what he perceived as five key types of alcoholism: alpha, beta, gamma, delta and epsilon. Alpha alcoholism was defined as 'a *purely* psychological *continual* dependence or reliance upon the effect of alcohol to relieve bodily or emotional pain' (p. 36; italics in the original). This form of drinking was held to occur contrary to the norms of society as they relate to alcohol consumption but did not lead to loss of control or an inability to abstain. Problems associated with such drinking were held to be restricted to interpersonal problems and nutritional deficiencies. This was not a progressive process and there would be no withdrawal symptoms. Beta alcoholism was seen to be consumption which led to certain alcohol-related medical disorders such as gastritis or cirrhosis of the liver. There would be no physical or psychological dependence or withdrawal symptoms. The negative consequences of such use would be nutritional deficiencies, impaired family budget, lowered productivity and a reduced life-span. Gamma alcoholism, Jellinek argued, led to: '1) acquired increased tissue tolerance to alcohol, 2) adaptive cell metabolism, 3) withdrawal symptoms and "craving" i.e., physical dependence, and 4) loss of control' (p. 37). For Jellinek this was the worst form of alcoholism in that the negative consequences present with alpha and beta alcoholism were present to a greater degree than with either of these other forms, and there was also loss of control. Delta alcoholism was defined as being largely the same as gamma alcoholism except that instead of loss of control there was an inability to abstain. Although the person depicted as a delta alcoholic may be unable to abstain from alcohol consumption for more than a short period, their actual drinking would not lead to loss of control on any specific occasion. Jellinek associated this form of drinking with wine drinking in France, where large daily amounts were consumed, which would lead to negative health consequences, but drunkenness, loss of control, was uncommon. Jellinek's final category, or species of disease as he termed them, was epsilon alcoholism. This entailed periodic alcoholism, where bouts of drinking could cause what he referred to as serious damage. Other more broad categories were described but Jellinek held that only these five patterns of use could be considered as possibly being a disease. He then in turn ruled out alpha and beta forms as they did not arise from a physical or psychological pathology. This is a crucial point because for alcoholism to be a disease there needs to be established a biological cause of excessive drinking. The fact that heavy drinking may lead to physical damage and consequent medical disorders was not sufficient. In consequent Jellinek regarded gamma and delta alcoholism as diseases as, he argued, they both involved: 'adaptation of cell metabolism, and acquired increased tissue tolerance and the withdrawal symptoms, which bring about "craving" and loss of control and ability to abstain' (p. 40). Epsilon alcoholism remained under consideration as it was felt that there was at the time insufficient knowledge about this form.

The essential difference between gamma and delta alcoholism was seen to arise from loss of control or an inability to abstain. Gamma alcoholism was held to lead to loss of control, whereas delta alcoholism was categorised by an inability to abstain. This difference arose from patterns of drinking determined by: 'social attitudes, the predominance of certain sources of alcohol and, in certain areas and social gradings, by economic factors' (p. 111). As Jellinek argued:

> The drinking pattern can determine the course of the alcoholic process, in the sense that such a pattern as described for France [the socially endorsed daily consumption of wine] can lead to a constant presence of alcohol in the organism with little manifestation of overt intoxication, and result in a species of addiction in which there is no 'loss of control' but instead an inability to 'go on the water wagon,' i.e., to abstain. On the other hand in the Anglo-Saxon countries there is – as the majority rule – no distribution of alcohol over the entire day but rather a shock like impact of strongly intoxicating amounts toward evening. This pattern can produce 'loss of control' but leaves

the ability to 'go on the water wagon,' to abstain for shorter or longer periods, practically intact.

(Jellinek 1960: 30)

Hence, according to this analysis, in Anglo-Saxon countries one could expect to encounter gamma alcoholism, whereas in France there would be a preponderance of delta alcoholism.

Nevertheless, despite the importance of social and cultural events in the genesis of patterns of drinking, it is important to note that, for Jellinek, both forms nevertheless had a biological cause. In other words, there is some latent characteristic in those who may display gamma or delta alcoholism such that certain social and cultural conditions will cause one or the other of these forms to occur. Although he hypothesised that:

In societies which have a low degree of acceptance of large daily amounts of alcohol, mainly those will be exposed to the risk of addiction who on account of high psychological vulnerability have an inducement to go against the social standards. But in societies which have an extremely high degree of acceptance of large daily alcohol consumption, the presence of any small vulnerability, whether psychological or physical, will suffice for exposure to the risk of addiction.

(Jellinek 1960: 28)

He also saw such psychological or sociological elements as being only: 'the preparation of the terrain on which addiction develops' (p. 72). Psychological and sociological antecedents to alcohol addiction were for Jellinek merely the starting point for a process whereby some individuals with certain physiological or biochemical anomalies (p. 155) would progress to levels of drinking which caused harmful physiological changes and thus the disease of alcohol addiction.

This was a sophisticated analysis which in many ways has yet to be bettered. Jellinek demonstrated clearly that alcohol addiction, by which he meant alcoholism as a disease (see p. 7) is a negative progression of alcohol use which has social, psychological and biological components. For it to be a disease he claimed that the latter must be present, but the other elements were also recognised as central elements in the process.

Present-day research has in many respects forgotten this important contribution and it is fairly common to read research reports and other academic publications which appear to portray alcohol and drug problems and misuse as though they were solely a social or psychological or biological concern. Jellinek's work should serve as a constant reminder that the picture is more complex. Each of these elements is an important part of any comprehensive account of intoxication. Alcohol and drug misuse may have social forms which change over time, but the experience and consequences are also embodied within a biological form. To demonstrate this latter point more clearly it is necessary to reflect briefly on current understanding of the biology of substance use and misuse.

NEUROBIOLOGY AND SUBSTANCE USE

Before proceeding with a discussion of the way in which current research illuminates the biological basis of drug and alcohol use, it is probably necessary first to present a brief description of some basic aspects of brain structure and chemistry. Technical details will be kept to a minimum, so those readers from a non-science background should not be deterred from engaging with this section.

Perhaps the first thing to grasp about the brain is how immensely complex it is in almost every respect. There are approximately 12 billion nerve cells in the average human brain, and each of these is linked to many others. It has been widely estimated that there are more intercellular connections in the human brain than there are individual particles in the whole universe. In addition the brain is subdivided into distinct morphological and functional sub-units and systems which are nevertheless interconnected in a complex way.

These interconnections between nerve cells within the brain are made via short fibres which are known as dendrites and long fibres called axons, both of which extend from the main body of the cell (each cell, or soma, has only one axon). Dendrites receive impulses from other cells and axons function to transmit impulses. Connections between dendrites and axons and other neuronal cells (the name given to the complete cell of soma and fibres) is mediated by gaps known as synapses which are bridged by transmitter chemicals released in response to the nerve impulse. Transmitter chemicals (neurotransmitters) can either excite or inhibit the receiving soma. Recent research has found that different types of neurotransmitters are found in differing regions of the brain and appear to be associated with varying brain functions. Those commonly associated with drug and alcohol use include serotonin (associated with mood, sleep, pain and tolerance to alcohol and other drugs), dopamine (motor functions, emotive arousal and tolerance to alcohol and other functions) and gamma aminobutyric acid (motor control and sensory processing).

Most current research relating to the biological basis of drug and alcohol use tends to focus on work relating to areas of the brain which, by common consent, are held to have evolved prior to those parts of the brain within which higher-order functions such as sensory response and cognition are generally agreed to arise (i.e. the thalamus and cortex). These more primitive areas are known as the brain stem and the limbic system. The brain stem comprises the medulla, pons and midbrain and is effectively an enlarged continuation of the spinal cord. It interconnects anterior brain structures (the thalamocortical system) and the spinal cord and is associated with emotive arousal, sleep and other sensory modes. The limbic system is an integrated network of structures including the hypothalamus which lie anterior to the midbrain and which are centrally involved in eating, drinking, sexual behaviour and emotive arousal. Edelman (1992) has depicted the brain stem and the limbic system as a single system which, through its functions in relation to the mediation of appetite, sexual and consummatory behaviour, serves to assess or depict value; by which he means survival advantage (see chapter three, this volume).

There are a number of events or experiences associated with alcohol or drug use which are often discussed in the literature as being evidence of a biological basis to substance use. The most commonly used terms are *craving* (a compulsion to continue use), *withdrawal* (an aversive affective state due to discontinued or interrupted use), *euphoria* and *tolerance* (reduced effects from exposure to alcohol or other drug). In addition to the use of these concepts, and in many respects overlapping with them, there is growing identification of what in psychology are referred to as positive and negative reinforcers. A positive reinforcer induces a desirable state which was absent prior to use, whereas a negative reinforcer lessens or extinguishes aversive states.

Ideas about drug and alcohol dependence within biological and medical science in the recent past, notwithstanding Jellinek's early work, largely tended to focus on concerns about withdrawal or negative reinforcement. Use continued so that aversive states such as those arising from physiological stress due to biological changes effected by the substance taken, or environmentally derived anxiety, stress or depression would be alleviated. More recently, however, researchers such as Wise (1988) have noted that positive reinforcement, the pursuit of euphoria, can be sufficient to produce desires to continue alcohol or drug use. Jaffe (1989) cites a considerable weight of evidence to illustrate that positive reinforcement can be grounded biologically through effects on certain neurotransmitters and is thus not merely a psychological property:

> A substantial body of evidence now points to dopaminergic neurons in the ventral-tegmental area [VTA] of the brain stem, their projections to the nucleus accumbens, and efferents from that nucleus to the ventral pallidum and thalamus as critical pathways for the reinforcing effects of cocaine and amphetamine. The reinforcing effects of many drugs seem to be due to their capacity to release dopamine or prolong its actions at the synapse by preventing its re-uptake. Because animals will readily self-administer these agents in the absence of any obvious distress, they are viewed as positive reinforcers. [Opioid agonists] such as morphine can also act as reinforcers at several points in this system. . . . These reinforcing effects can be demonstrated in animals that are not physically dependent.
>
> (Jaffe 1989: 52)

Jaffe further states that other substances, such as alcohol, barbiturates, nicotine and PCP can also have some capacity to activate the ascending dopaminergic system.

These findings are supported by Bozarth (1994) who, in discussing opiate use, argues that both positive and negative reinforcement are key components in drug use. He cites evidence which gives further support to the growing acceptance of the importance of dopamine systems in the ventral tegmental area of the brain as a mechanism for positive reinforcement. Holman (1994) attests to evidence which illustrates that dopamine and the ventral tegmental area may also be central to the positive reinforcement of stimulants. Hence this system would appear to have a generalised response to a range of drugs of misuse. Indeed, Balfour (1994) has noted the importance of the DA-VTA in the self-administration of nicotine, and Littleton and Little (1994) have argued that the same mechanism is a site for the positive reinforcement of alcohol in terms of stimulant and euphoric effects. However, Holman (1994) also points out that DA agonists do not prevent continued drug misuse, so the DA-VTA may not be the site of a primary effect but merely a regulator mechanism. And Holman (1994) has made an important contribution when noting that drug dependence and relapse are not effected by the DA-VTA mechanism. It would thus seem that different brain areas and neurotransmitters appear to be involved in drug or alcohol withdrawal than are involved in positive reinforcement.

Jaffe (1989) noted such differences with respect to opioids in that: 'the neural systems critical to their positive reinforcing actions are anatomically and biochemically distinct from those which are involved in the more dramatic components of the opioid withdrawal syndrome' (p. 53). According to Balfour (1994), the desensitisation of receptors associated with serotonin (which, you will recall, is linked to mood and sleep) is prominent in nicotine use in terms of negative reinforcement. Lader (1994) makes a similar point with respect to benzodiazepine use. And Holman (1994) records that the neurotransmitters or receptors serotonin, acetylcholine (a widespread neurotransmitter which elicits a response in receiving cells) Gamma-aminobutryic acid (GABA) (inhibits receiving cells and, as noted above, is linked to sensory processing and movement) and N-methyl-D-asparate (NMDA) a receptor which may contribute to alcohol withdrawal seizures (Hiller-Sturmhoefel *et al.* 1995) may all be linked to various experiences of negative reinforcement.

What these findings illustrate is that the sites and mechanisms which effect or promote the positive or negative experiences of alcohol or other drug use may be quite distinct. Processes which lead to continued self-administration may be different from those which lead to what is traditionally known as dependence, that is a biologically grounded need to continue use to avoid aversive reactions such as withdrawal. One of the key findings of this research is that positive reinforcement can now be seen to arise in part from biological foundations, and is not just a matter of psychological predisposition. But as important as this conclusion is, it is still unclear which of positive or negative reinforcement is most influential with respect to continued self-administration. Do people primarily continue to use because it gives them pleasure or because they are avoiding unpleasant consequences of stopping use? The picture is complex according to Jaffe (1989) as the level of a withdrawal experience *per se* cannot reliably predict whether self-administration will continue. He notes that the relatively mild withdrawal experience of nicotine appears to motivate continued use in that nicotine gum, which has only low positive reinforcing effects, can facilitate the cessation of tobacco smoking, whereas the serious withdrawal effects of phenobarbital are not sufficient to promote self-administration.

This complexity probably signals the fact that any drug-using experience arises, at the level of brain chemistry, from the specific effects of given substances upon particular neurobiological sites. This in turn leads to a diverse modulation of emotive states and thus a substance-specific balance of positive to negative reinforcement. The experience of, say, taking cocaine may arise from a different positive to negative configuration than that, say, of benzodiazepines, but each of the two categories of reinforcement will be represented in the total event. Indeed, it would seem that established distinctions between stimulants and depressants may present a false dichotomy. A more accurate picture may reside in the possibility that the experience of a given substance and its power to engender self-administration, may arise at the biological level from the complex resolution of neural responses which, in turn, mediates the genesis of a composite reinforcing event at the level of the individual. But more of this later; at this stage it is enough to recognise the similarities in biological mechanisms between substances, as noted by Grunberg (1994), and the importance of both positive and negative reinforcement in sustaining use.

In a sense, if these processes occur at the biological level then they are also genetic, at least to some degree. It must be clear that whatever part the environment plays in effecting drug and alcohol use, some part of that predisposition must be inherited from our parents, from their parents and onwards to our distant ancestors. Of course the question is crucial, as any observation of drug taking and drinking practices will readily reveal that patterns of use vary considerably between individuals. The extent to which these differences are inherited is a key question which attracts considerable attention, and this will be looked at in the next section.

GENES AND INHERITANCE

It is commonly appreciated that many human traits such as hair or eye colour or height are inherited from a person's parents; that they run in families. The acceptance that many physical traits are inherited is indeed not new, but increasingly evidence is being presented which demonstrates that an individual's genetic endowment (their genotype) can affect their physical form (i.e. their phenotype) in less observable ways, such as with regard to a person's medical status. For example, there is evidence that a predisposition towards developing heart disease or certain forms of cancer are, to some extent, inherited. And recent genetic research has identified genes which are involved in the development of a wide range of diseases, including cystic fibrosis, the early onset of breast and ovarian cancer, spinal muscular atrophy and Alzheimer's disease. In some cases there is clear evidence that a genetic predisposition is closely affected by environmental factors, such as in the case of heart disease where diet may be centrally involved, whereas in others, for example Down's syndrome, it seems to be less so. Height is another example of the interaction between genotype and environment in determining a person's phenotype. As with heart disease, for most people the height to which they will grow arises jointly from both their genotype and diet.

The genetic information which comprises a person's genotype is encoded in a long chemical chain known as DNA (deoxyribonucleic acid). DNA is made up of nucleotides; a sugar molecule and a phosphate group attached to one of the four organic bases adenine (represented by a capital A), cytosine (C), guanine (G) and thymine (T). A combination of three nucleotides codes for one specific amino acid, these amino acids being the building blocks of proteins. For example, the nucleotide sequence TGG codes for the amino acid tryptophan, GAG for glutamic acid and AAG for lysine. In consequence, the order of nucleotides on a section of DNA will code for a sequence of amino acids which will in turn give rise to a specific protein; different sequences produce different proteins. Hence it is the sequence of nucleotides on DNA which is termed the genetic code; the section of nucleotides on DNA which codes for a single protein is termed a gene. Proteins are central to biological development because specific types form into enzymes which catalyse the generation of biological tissue.

So far, then, it has become apparent how the structure and form of biological tissue can be related back to a genetic code which resides in human cells. However, the question remains as to how this code can be passed on during reproductive processes to form new offspring. Each strand of DNA is arranged within chromosomes which are located in cell nuclei. In human beings there are

46 chromosomes arranged in 23 pairs. In women there are 23 homologous pairs, whereas in men there are 22 homologous pairs and one pair which consists of an X and a Y chromosomes (women have an XX pair). These are known as the sex chromosomes as they give rise to sex differences. Chromosome pairs have different lengths and sizes, which means that pairs can be discerned one from another.

During the formation of sperm and ova the number of chromosomes halves so that each resultant cell (known as a gamete) has only 23 single chromosomes. The number 23 is termed the haploid number of chromosomes. At fertilisation a gamete from the female and one from the male unite to form a fertilised cell called a zygote which, thus, once again, has 23 pairs of chromosomes. The resulting 46 chromosomes are termed the diploid number, which is a mixture of chromosomes from both parents.

The next step in this short journey into genetics requires some reflection on the work of Mendel, a monk born in 1822, who, in his ground-breaking research on peas, found that when he crossed plants with a smooth seed shape with one that had a wrinkled shape only smooth shapes resulted. However, when these plants self-fertilised only 75 per cent of the resulting seeds were smooth and the remaining 25 per cent were wrinkled. Mendel (1865) argued that this outcome arose as a result of the action of factors which could be dominant, in the case of smoothness, or recessive in the case of wrinkling. The process works in the following way. A smooth seed plant that only ever produces smooth seeds even after self-fertilisation will have only dominant factors, we can write this as SS to depict dominance in both male and female cells. A wrinkled seed plant which only ever produces wrinkled seeds after self-fertilisation can be shown as ss to depict recessiveness. Now, if SS is crossed with ss the only pairing which is possible is Ss; a dominant factor from one parent pairing with a recessive from another. However, if during the next stage Ss is crossed with Ss then 25 per cent of the resultant seeds will be SS (smooth) 50 per cent Ss (smooth as the dominant factor will be expressed) and 25 per cent ss (wrinkled as only recessive factors are present). The overall ratio would thus be 3 : 1 smooth to wrinkled, which is what Mendel found. Table 1 shows the results of the cross between Ss depicted in the vertical column with Ss depicted in the horizontal column. As is shown, the result is 25 per cent SS, 2 × 25 per cent Ss and 25 per cent ss.

Table 1 Results of cross between
Ss pea strains

	S	s
S	SS	Ss
s	Ss	ss

At the time Mendel did not have the benefit of contemporary understanding of molecular genetics, hence his use of the term 'factor'. If we translate these findings into what is now known about DNA and genes the picture should become somewhat clearer. As discussed above, the code which determines the structure of proteins from which arises all other tissue generation is held within DNA, and DNA is in turn held within chromosomes which are paired. Now genes, which code for a specific protein, occupy a position on a chromosome which is termed a locus. It is a gene locus which Mendel referred to as a factor, and from his work we know that a locus can have one of two forms, either dominant or recessive. These alternate forms of a gene locus are known as alleles and one allele is found on each of paired chromosomes.

From this it can be readily appreciated that sexual reproduction, being based on the mixing of chromosomes from each parent, can give rise to genetic changes which have consequences for an organism's phenotype, as in the case of smooth or wrinkled seeds. If added to this is the possibility of mutation, where the gene sequence can be changed such that a protein slightly different from the norm can arise, and recombination, where genes on the same chromosome recombine with each other during gamete formation, then it should be clear that genetic changes can have large-scale implications for a person's phenotype, or physical/biological form.

By now you may be asking what all this has to do with alcohol and drug use. As demonstrated above, there can be little doubt that alcohol and drug use is grounded in part in biological processes. Now it should also be clear that biological processes can be influenced greatly by factors which govern the inheritance of a genetic code from which biological development takes place. Given this, it should be possible to show to some degree to what extent and in what way patterns of alcohol and drug use are inherited.

THE INHERITANCE OF ALCOHOL AND DRUG USE

It is not difficult to show that alcohol problems tend to recur in families. Cotton (1979) provided a review of research carried out on family drinking problems which had been undertaken over a forty-year period and involved some thirty-nine separate studies. The main finding of this review was that irrespective of the sample being studied, alcoholics (so-called) were more likely than non-alcoholics to have an alcoholic relative. Of course, even though alcohol problems may run in families this does not necessarily mean that they are genetic in origin, they may instead arise from learned behaviour. Much current research seeks to illuminate the role of nature and nurture in drug and alcohol problems, through the application of a variety of analytic approaches. The intention in this section is to provide a concise overview of some of the key directions this work is taking. Hence there is no attempt here to be comprehensive. Readers with a particular interest in this area are advised to refer to the many excellent reviews which have appeared recently (e.g. Goodwin 1989, Anthenelli and Schuckit 1991, Cook and Gurling 1991, Crabbe and Goldman 1992, Heath 1995).

ADOPTION STUDIES

Perhaps the most widely known research relating to the inheritance of alcohol or drug problems is that concerning adoption studies. As Goodwin (1989) has noted, the first study of alcohol use which focused on adopted siblings was carried out in 1944 by Roe. This work found no difference in alcohol use between young people in their early twenties who were either offspring of 'alcoholics' or 'non-alcoholics'. The first evidence for a genetic link on the basis of adoption studies was put forward by Goodwin and co-workers in the 1970s (Goodwin *et al.* 1973, 1974, 1977, reported in Heath 1995). Through the use of official registries this group compared the children of people with a history of alcohol problems who were given up for adoption with a control group of such children of people with no such history. They also identified a sub-group of parents who raised one child but had a further child adopted. The results estimated a risk ratio of 3.6 : 1 for the adopted-away sons and 3.4 : 1 for the non-adopted-away sons of people with alcohol problems when compared with the same groups of children from the control group. The data did not show the same estimated genetic influence for women.

Later studies in Sweden by Cloninger and co-workers (discussed in Heath 1995) also demonstrated positive risk ratios for the sons of fathers with alcohol problems who were given for adoption when compared to a control group (1.3 : 1) and similarly the adoptee daughters of mothers with alcohol problems when compared with a control group (2.9 : 1) (Cloninger *et al.* 1985). Cadoret (1994), reporting on studies completed in the 1980s which were based on data from Iowa from the Lutheran Social Services and the Iowa Children and Families Services, also showed that sons from parents with alcoholic problems were approximately 3.5 times more likely to develop alcohol problems than a control sample. The data from the Lutheran Social Services study also showed an increased risk for women who were adopted, but later research using the Catholic Adoption Agencies (Cadoret 1994) estimated a reduced risk (see Heath 1995).

TWIN STUDIES

A further form of research has involved studies of the risk of alcohol problems for the twins of those defined as alcoholics (although it should be noted that there is no standardisation of this term between different studies). The studies compare the risk of developing alcohol problems for identical, monozygotic twins (from the same zygote) with fraternal, dizygotic (from twinned zygotes). Monozygotic twins share the same genes, whereas dizygotic twins share on average only 50 per cent of their genes, hence different rates of inheritance between the two groups should be found if genetic influences are important in the development of alcohol problems.

Goodwin (1989) noted that studies of fraternal and identical twin drinking practices, alcohol metabolism and the effects of alcohol on brain waves as measured by electroencephalogram (EEG) all reported high degrees of inheritability. Twin studies have also shown a genetic influence with respect to alcohol problems, or alcoholism as it is usually referred to in this literature. Heath (1995) stated that work by Kaij (1960) found that rates of alcohol problems coexisting between monozygotic twins were higher than between dizygotic twins. Similar results were found by Hrubec and Omenn (1981), Koskenvuo *et al.* (1984), Romanov *et al.* (1991)

and Allgulander *et al.* (1991, 1992). For a detailed discussion of these findings the reader is referred to the excellent review and re-analysis of this work carried out by Heath (1995) who concludes that 'The reanalysis . . . has confirmed the consistency of the evidence for an important genetic influence on alcoholism risk from both twin and adoption studies' (p. 170). Heath (1995) also notes that results have been found to be consistent across time periods, to apply both to men and to women, to be consistent despite variance between samples with respect to definitions of alcohol problems and to display cross-cultural differences with respect to the contribution of genetic influences to the risk of developing alcohol problems. In contrast, Goodwin (1989) records that findings by Partanen *et al.* (1966) found such effects to be present with respect to young twins but not older twin pairs, and that Murray *et al.* (1983) found no difference between fraternal and identical twins.

MOLECULAR APPROACHES

In addition to the above work on twins and other siblings, further research in genetics has also demonstrated that nature has a part to play in the emergence of alcohol and drug problems. As was stated above, it is not the intention in this text to provide a full and detailed discussion of the range and findings of current research, but instead to provide for the reader with little knowledge of this area insights into current work on the genetics of alcohol and drug use so that they can follow subsequent discussion of substance problems as a complex phenomenon. However, notwithstanding this caveat, two more areas of research are worthy of brief discussion; the search for biochemical markers and the analysis of quantitative trait loci.

The study of biochemical markers is directed towards discovering which individuals may be a greater risk of developing alcohol or drug problems. To date two specific markers have been found to be linked to the metabolism of alcohol; monoamine oxidase (MAO) and adenylyl cyclase (AC). MAO is an enzyme that functions to break down monoamine neurotransmitters such as dopamine and serotonin. As discussed above, these transmitters have been found to be associated with mediating the effects of both alcohol and drug use. Research has shown that low MAO activity may be associated with alcohol problems that begin at an early age (defined as Type II alcoholism by Cloninger 1987) through altering the breakdown of these specific neurotransmitters (Anthenelli and Tabakoff 1995). These authors also noted that irrespective of classifications of alcoholism 'men with an earlier age at onset and a more severe course of alcohol related problems had significantly lower platelet MAO levels than nonalcoholic men' (p. 177). Sher *et al.* (1994) carried out a review of the evidence that low MAO levels are inherited and reported that some studies show inheritability whereas others do not. Anthenelli and Tabakoff (1995) record that what evidence there is appears to relate to early onset, Type II alcoholism.

Adenylyl cyclase relays signals from the exterior to the interior of neural cells. Tabakoff *et al.* (1988) found that AC activity was lower in people with alcohol problems than those without. Anthenelli and Tabakoff (1995) report that these findings remained even when controlling for age, race, smoking and use of other drugs. Consequently, these findings appear to suggest that AC activity may be an inherited characteristic which imparts a predisposition to alcohol problems. However, the evidence for this has yet to be fully established. It appears that AC levels can be influenced directly by alcohol consumption as people with alcohol problems who were current drinkers were found by Waltman *et al.* (1993) to have the same levels of AC activity as a control group, whereas abstinent alcoholics had lower AC activity. Anthenelli and Tabakoff conclude that 'the usefulness of AC activity as a marker for alcoholism remains somewhat controversial' (p. 179). It appears from recent research that AC activity is modulated by an inhibitory guanine nucleotide-binding protein (G protein) (Anthenelli and Tabakoff 1995) and that one gene may lead to levels of G protein (Devor *et al.* 1991, quoted in Anthenelli and Tabakoff 1995). The evidence appears complex but on balance it does seem to be the case that both MAO and AC activity affect the way in which alcohol is responded to biochemically, and that differences in activity may, at least in part, be inherited.

So far this research refers to the effects of single genes on alcohol or drug use, but there is some increasing evidence that there are quantitative differences between people with respect to certain traits associated with drug or alcohol use. That is, it is not solely a matter of whether a trait is present or absent, but rather the degree to which it is held, as in the case of height. It was pointed out above that the site of a single gene is termed a locus and, similarly, the site on the genome of a quantitative trait is termed a quantitative trait locus (QTL). A specific quantitative trait can result from the combined action of a number of different QTLs in different parts of the genome. For the purposes of this analysis it is not necessary for the reader to understand the methods used to map QTLs, but those who do require further information are referred to Plomin *et al.* (1991), Gora-Maslak *et al.* (1991) and Grisel and Crabbe (1995).

Recent work with mice which have been inbred to produce strains with genetically identical members has shown QTLs to effect alcohol-induced hypothermia, locomotor activity, withdrawal and the acceptance of alcohol solutions (see Crabbe and Goldman 1992, Grisel and Crabbe 1995). Crabbe *et al.* (1994) (cited in Grisel and Crabbe 1995) found two markers on chromosome 9 associated with alcohol-induced hypothermia which were in the same region as the gene which coded for serotonin. As discussed above, this neurotransmitter is closely associated with the effect of alcohol and other drugs. A further QTL has been found close to the gene which codes for part of the acetylcholine receptor. As reported in Grisel and Crabbe (1995), Rodriguez *et al.* (1995) have stated that this QTL is associated with a number of alcohol-related responses, including alcohol acceptance and motor activity. And further research findings show that several QTLs may influence alcohol withdrawal symptoms. Crabbe *et al.* (1994) found that a specific QTL accounted for 40 per cent of the genetic basis of alcohol withdrawal (quoted in Grisel and Crabbe 1995).

THE BIOLOGY OF ALCOHOL AND DRUG USE

The evidence presented in this chapter makes it difficult to draw any conclusion other than that it is clearly the case that drug and alcohol use has a biological basis and that certain traits are inherited. In many ways, as recorded at the beginning of this chapter, this conclusion is in some ways inevitable. We are a biological species and thus it should come as no surprise if we find that things that often give us pleasure are in part grounded in our biological form. The range of evidence is impressive. Current research has been able to show the extent to which both positive and negative reinforcement, terms which are more common in psychology than biology, may arise from distinct aspects of brain biochemistry. That is, how certain neurotransmitters, acting in distinct areas of the brain, mediate the effects of alcohol and other drugs to give rise to positive and negative experiences which in turn promote or inhibit consequent patterns of use. The fact that terms such as craving, withdrawal or euphoria can be seen to be grounded biochemically and physiologically is highly important. It is not uncommon in many texts on the subject of craving for, say cocaine, for it to be described as a psychological addiction without a biological basis. The research findings discussed here offer evidence contrary to that view. The idea that a compulsion to use which overrides social medical or economic well-being can only be understood in terms of negative reinforcement, the avoidance of withdrawal, is now unsustainable in face of the evidence now being collated by many research teams. The role of neurotransmitters, which act in the lower areas of the brain which are more primitive in evolutionary terms, has been found to be central to positive reinforcement and the experience of euphoria through drug and alcohol use. Evidence that positive reinforcement is important in sustaining self-administration is important in this regard. It has been noted above that individuals will not continue to administer drugs which have beneficial effects on their health in the absence of positive reinforcement, whereas the opposite is true of extremely harmful substances such as tobacco.

Of course this is not to say that negative reinforcement is not important. Research has shown that this too acts to sustain use, and that distinct biological mechanisms have been proposed as the means through which some observed drug- and alcohol-related behaviour is given effect. What current findings do show most clearly is that both positive and negative effects of drug and alcohol use need to be considered in any comprehensive analysis of biological mechanisms. And that these separate consequences of use arise from different neurobiological bases.

Even though it can be established that alcohol and other intoxicants act by way of biological pathways this is not to say that such use does not in the final analysis arise also from environmental, social or cultural antecedents. It could be argued that the use of alcohol, cocaine, heroin or some other substance in fact changes brain biochemistry such that problems of use arise. By chance person A uses more than person B and therefore in the course of time person A, through biochemical changes effected by their use, develops problematic use. That is, the physiological and biochemical differences seen between those that experience alcohol and drug problems are induced by their use, which originated otherwise in social and cultural contexts. It is this issue – inheritance or environment – which has featured strongly in the agenda of many alcohol and drug researchers.

The evidence presented above demonstrates clearly that physiological and biochemical responses to alcohol and drug use are at least in part inherited. A wide range of evidence, from adoption and twin studies to the identification of biological markers for responses to drug and alcohol use, have supported the findings that certain traits associated with use are inherited. It appears from the evidence that differences in the way alcohol or other drugs are metabolised, or perhaps even experienced, can give rise to dissimilarities between levels of use. We have also seen that these differences may not only be qualitative in nature (i.e. either present or absent) but may vary quantitatively between individuals. Hence these findings allow for large distinctions to be found between users with respect to levels and patterns of use. Nevertheless, as important as these findings are they do not support a belief in the sole role of biology in engendering substance problems. As Jellinek (1960) argued, cultural differences play an important part in the way alcohol problems arise. He stated that drinking outcomes for the individual can be affected by whether a culture is tolerant or intolerant towards specific forms of drinking. Later research, including both twin and adoption studies, has shown enhanced rates of alcohol problems for the children or siblings of people with alcohol problems but it did not show that such problems were inherited in every case and to the same extent. As Crabbe and Goldman (1992) pointed out: 'As children growing up in alcoholic households have an increased risk of becoming either alcoholic or abstinent, it seems that increased risk of alcoholism depends partly on how a person *reacts* to his or her environment' (p. 299; italics in the original). Hence it would seem that a person's genotype can be transcended or its effects can be modified in certain circumstances, and thus outcomes different from those predicted by their biological form can arise. How this may happen, how external circumstances can effect changes in the way we are, is the subject of the next chapter.

Chapter 2

Biology and experience

The central message of the previous chapter is fairly clear; whatever else drug and alcohol use and abuse may be, it cannot be disputed that these practices are embodied in the sense that the experience is grounded biologically and that certain traits are inherited. However, as the evidence has shown, this is only one part of a more complex whole. The fact that the incidence of alcohol-related problems amongst certain groups of twins may occur at a rate 3.5 times higher than the general population is an important finding, but caution is needed in the interpretation of such statistics. Although this can be used correctly as evidence for the heritability of alcohol problems, the data also allow for the possibility that the environment may play as great or even a greater role in determining individual outcomes. The finding by Valliant (1983) that children who grow up in households with an alcoholic parent are as a group on average as likely to become teetotal as develop alcohol problems is important in this respect. As noted in Chapter 1, Crabbe and Goldman (1992) concluded that: 'As children growing up in alcoholic households have an increased risk of becoming either alcoholic or abstinent, it seems that increased risk of alcoholism depends partly on how a person *reacts* to his or her environment' (p. 299; italics in the original). What we really need to know is what proportion of a person's actions stem from their biological self prior to any external input, and what arises in consequence of experience and the environment. In other words, how much is nature and how much nurture?

GENETICS, ENVIRONMENT AND EXPERIENCE

Although it is of interest to discover and compare rates of alcohol problems amongst twins compared with those to be found amongst a general population, what we really want to know is what proportion of these differences can be attributed to genetic as opposed to environmental differences. One way of doing this is to calculate a heritability statistic, showing the proportion of phenotypic difference which arises from genetic effects. A heritability statistic is a modified regression analysis. A regression analysis can be used to depict the way in which one trait varies in relation to another (say, height and weight) or how a specific trait varies (correlates) between related individuals. It gives a correlation coefficient or regression slope the size of which indicates the degree of heritability. To arrive at a heritability statistic the slope of the regression or correlation coefficient is modified according to genetic relatedness. For a regression of offspring on one parent the heritability statistic would be the correlation coefficient divided by 0.5 (the proportion of shared genes), whereas for a regression on both parents it would be the coefficient divided by 1.0 (both parents account for all the genes). This is a rather simplistic description of a complex procedure, but it will suffice for the purposes of this present text. Readers interested in gaining greater understanding of this complex area are referred to Loehlin (1992) or Neale and Cardon (1992).

Heritability statistics can be used to show the genetic contribution to a wide range of traits or factors. For example, Plomin (1994) reports the correlations in height of .90 and .45 for identical and fraternal twins respectively give rise to a heritability estimate of 90 per cent (p. 44). As Plomin has stated heritability is:

> an estimate of effect size given a particular mix of existing genetic and environmental factors in a particular population at a particular time. It is a descriptive statistic that estimates the proportion of phenotypic variance (i.e., individual differences in a population, not behaviour of a single individual) that can be accounted for by genetic variance.
>
> (Plomin 1994: 43)

Heritability statistics have been used widely in studies of the genetic basis of human behaviour, particularly with twin studies. For example, Plomin *et al.* (1988), in looking at adult twins' retrospective ratings of their childhood family environment found heritability statistics of 0.24 for expressiveness, 0.32 for conflict, 0.35 for achievement and 0.31 for cultural factors. In reviewing a wide

range of evidence relating to the genetic basis of children's perceptions of the family environment, Plomin (1994) concluded that:
'Evidence for genetic contributions emerges for all dimensions of
children's perceptions of parenting with the interesting exception
of control-related criteria. Children's perceptions of their siblings'
behaviour toward them also shows genetic effects' (p. 79). In addition to the above, Plomin (1994) also reviewed evidence relating
to non-family environment. In citing data from the Non-shared
Environment in Adolescent Development project (NEAD),
Plomin (1994) reports high levels of sibling heritability of choice
of peers as calculated from parental ratings of siblings' peers. In
fact heritability statistics of peer substance abuse were found to
be 0.72 from mothers' data and 0.74 from fathers' data. Many
studies have also been carried out on life events (Holmes 1979).
Items used in calculating life events include marital, financial,
health and employment difficulties. This research investigates
whether or not certain events happen to anyone, or whether unfortunate occurrences happen only to some people. In terms of engendering disadvantages which may lead to alcohol or substance
problems this type of research is of some interest to the main subject
of this current text. Overall, data show that life events have an estimated heritability of 0.31 (Plomin 1994). Genetic influences were
also found to exist with respect to friends, teachers, social support,
accidents, classroom environment, education and socioeconomic
status. With respect to the latter, findings from a large number of
studies (Fulker and Eysenck 1979, Taubman 1976, Teasdale 1979,
Teasdale and Owen 1981, Tambs *et al.* 1989, Lichtenstein and
Pedersen 1991) indicate a heritability statistic of about 0.40 for
occupational status (cited in Plomin 1994).

Clearly these are important findings for if it can be shown that
environmental factors such as occupational status or selection of
friends has a genetic component then it cannot be assumed that
what happens to you throughout your life is solely and completely
a social or cultural matter. However, the issue cannot be left there
as it is important to establish what mediates genes and environments. That is, what trait or traits, being genetically determined,
give rise to specific environmental outcomes? What we need to
know is how genes and environment are linked. Plomin *et al.*
(1977) described three types of correlation between genotype and
the environment. The first is termed 'passive' and refers to a correlation which arises from shared heredity and home environment.

The second genotype environment correlation is termed 'reactive' in that experiences derive from the reactions of others to a person's genetic disposition. The third is an 'active' correlation in that a genetic disposition leads to the selection of particular environments. If, by way of illustration, we consider these possibilities in terms of the mediation of drug or alcohol use then passive correlation would be where problematic drug or alcohol use was facilitated in a drug- or alcohol-using family environment by parents who were also pre-disposed genetically to such problems. A reactive correlation may be where a person genetically disposed towards problematic use developed such problems in response to opportunities provided by friends or associates. Third, active correlations would indicate the selection of problematic alcohol- and drug-using circumstances by those genetically predisposed to problematic use.

The analysis of such genetic-environmental links is carried out through the multivariate analysis of covariance between behavioural or psychological traits and environmental measures. Through the use of multivariate analyses the extent to which the genetic effects on, say, a behavioural trait are shared with those on an environmental measure can be assessed. That is, the degree to which behavioural traits and environment measures are mediated by genotype can be determined.

These techniques have been used to examine a variety of phenotypic and environmental measures. For example, data from the Swedish Adoption/Twin Study of Aging (SATSA) (Bergeman *et al.* 1991) has been used to analyse the relationship between social support, depression and life satisfaction. Plomin *et al.* (1991) found evidence to suggest that the genetic effects on social support can be accounted for in part by genetic effects on depression and life satisfaction (reported in Plomin 1994). Hence it cannot be assumed that, as may be supposed, depression acts causally to decrease both life satisfaction and social support. The evidence seems to suggest that the association between these phenotypic and environmental elements is in part genetically mediated. Similar analyses have been carried out with respect to life events and personality (Brett *et al.* 1990, Neiderheiser *et al.* 1992), socioeconomic status and intelligence (Tambs *et al.* 1989, Lichtenstein *et al.* 1992). In each case these paired measures were found to be mediated genetically to varying degrees.

These findings are important, but caution needs to be exercised in interpreting this research too favourably at the present time in terms of establishing the primacy of genes over the environment. As Plomin (1994) has noted: 'A general hypothesis is beginning to emerge from multivariate genetic analysis of this type: Phenotypic covariance between an environmental measure and traditional trait measures is typical due in part, *but only in part*, to genetic mediation' (p. 148; italics added). And in terms of types of genetic–environment correlation Plomin states that:

> taking the data at face value, it appears that passive GE [genetic-environmental] correlation is a stronger force than reactive and active GE correlation in terms of mediating the genetic contribution to environmental measures . . . a developmental theory of genetics and experience predicts that the reactive and active forms of GE correlation become more important as children experience environments outside the family and begin to play a more active role in the selection and creation of their environments.
>
> (Plomin 1994: 148)

Drawing together the results of many years research on the genetics of behaviour Plomin (1994) has constructed an explicit theory of the relationship between genetics and experience. Some aspects of this theory are supported by ongoing research findings whilst others are speculative at this stage. His work clearly demonstrates the potential of current research to show clear links between genetics and experience, and warrants careful consideration. Consequently the seven hypotheses which underlie his position need to be considered.

Plomin (1994) has outlined seven hypotheses which he argues represent an empirically based theory of genetics and experience. These hypotheses are as follows:

1 Genetic differences among individuals contribute to measures of the environment (p. 160). A variety of findings including those cited above indicate that some environmental measures, such as socioeconomic status, have been found to have a genetic component. However, it is not claimed that all environmental measures have a genetic component, nor that all the variance in environmental measures that do show a genetic effect is genetically grounded. The importance of these findings stems from the fact that factors which may be thought to be solely non-genetic in origin may in fact be in part genetically grounded.

2 The genetic contribution to measures of the environment is greater for measures of active experience (p. 161). Plomin's hypothesis is that environmental measures for which the individual displays active choice will show the greatest heritability. To support this hypothesis it is claimed that data on life events show a positive relationship between level of heritability and degree of personal control. In some ways this is, at this stage of research findings, a rather weak claim. Data on twin correlations and heritabilities for adult twins' ratings of life events quoted by Plomin (p. 93) show heritability statistics for marital difficulties (0.14), interpersonal difficulties (0.18), illness/injury (0.21), being robbed or assaulted (0.33) and financial problems (0.39). It is difficult to determine in any absolute way which of these categories is subject to greater personal control than any of the others. Indeed one could argue for the existence of greater control over interpersonal relations than a likelihood to be robbed or assaulted. Nevertheless, despite the current lack of evidence for the hypothesis, which Plomin acknowledges, this may prove to be an important area of research. Given that the heritability statistic is a measure of variance in a population, then where there is no variance there will be no measurable inheritance in the sense implied here. Consequently, any genetic characteristic which is required for the survival of an individual (increases genetic fitness) will show little variation in a given population, otherwise survival chances will decrease. Conversely, those characteristics which are less central to survival can tolerate greater variance. In this sense greater variance means greater 'choice' on the part of the individual and hence greater heritability may be expected. The fact that we may see inheritance through choice is an important insight which will be returned to below.

3 The genetic contribution to measures of the environment is due in part to psychological traits (p. 161). This proposition suggests that the covariance to be found between environmental measures and phenotypic traits such as cognitive abilities or psychopathological dysfunction may have genetic components. As discussed above, such effects have been found with respect to life events and personality, socioeconomic status and intelligence, and socioeconomic status and health. An important consequence of this proposition is that the link between genotype and certain environmental measures may be mediated by a phenotypic trait or traits. Hence, if I can be permitted some creative speculation for a moment, it may be that an association between an alcohol-related gene and a problematic alcohol consumption outcome could be mediated by a phenotypic trait which promotes the seeking out of alcohol-based social circumstances. We could call this trait, say, 'risk taking'. However, we need once again to recall that, as Plomin frequently states, whatever role a person's genotype plays, it is only part of the equation.

4 Genetic differences among individuals contribute to differences in experience independent of psychological traits (p. 162). It may be that the genetic effects on some measures of the environment are not mediated by a psychological trait. It is suggested that this may be difficult to demonstrate as it may always be possible to argue that some as yet undiscovered X factor could be involved. However, the hypothesis is important in that it allows for the possibility that the way individuals interact with their environment is influenced directly by genetic factors. This is an exciting possibility as it means that the form experience takes is genetically derived, at least in part. This complements Edelman's ideas about the role of natural selection in determining patterns of behaviour through the way experience gives rise to particular neural pathways in the brain. These important ideas will be addressed directly in chapter three.

5 Genetic factors contribute to links between environmental measures and developmental outcomes (p. 163). Both environmental and developmental outcome measures show genetic effects. In consequence Plomin argues that the link between environmental measures and outcomes may be genetically mediated. For example, it may prove to be the case that correlations between, say, life events and subsequent depression may be explained in part by the existence of genetic effects on both the environmental measure and the outcome. That is, the covariance between environment and outcome will partly arise from shared genetic effects.

6 Processes underlying genetic contributions to experience change during development (p. 163). As already discussed, during developmental processes there appears to be a shift from passive correlations between genotype and environment to active correlations. To understand this shift it is important to develop greater understanding of children's acquisition of experience. As Plomin argues: 'Progress in understanding this shift from passive to active GE correlation – indeed, the progress for the entire field of genetics and experience – depends on developing measures of

children's active selection, modification, and creation of experience'
(p. 163). Once again these reflections are consistent with Edelman's
theories of the evolution of the mind which also focus on the
importance of selection and modification in the development of
experience.

7 Specific genes that affect experience will be identified (p. 163).
Plomin states that as yet there is no evidence to support this
hypothesis, but notes that it is his belief that advances in molecular
genetics will eventually provide such findings. He notes that over
recent time research has identified many genes which may account
for the genetic variance in complex traits (Plomin 1990, 1993). The
reader will also recall from chapter one that considerable research
effort has been directed towards establishing which genes directly
affect drug and alcohol use. As yet, though, conclusive evidence
of such direct genetic effects is not available.

Plomin argues that: 'The long-term goal is to identify a set of DNA
markers of genes that accounts for a substantial portion of the
genetic variance for a particular measure of the environment'
(1994: 164). In consequence of these developments he further
states that the first six hypotheses will be rewritten to focus on
specific genes rather than genetic effects on covariance. The hypotheses are thus given as:

1 Specific genes will be identified that are associated with measures of the environment.
2 These genes are most likely to be found in association with
   measures of active experience rather than passive environments.
3 These genes will in part be associated with psychological traits.
4 These genes will in part be independent of psychological traits.
5 Genes associated with environmental measures will also be
   associated with outcome measures.
6 Genes will be identified that are associated with passive aspects
   of the environment in childhood, but later in development
   genes will be increasingly associated with active experience.

(1994: 164)

These hypotheses are important in that they make clear the agenda
of researchers seeking understanding of the role of inheritance in
effecting individual environmental outcomes. And they also
mirror the forms of research currently being undertaken by those

biologists and geneticists whose area of research is alcohol and illicit drug use. At the end of chapter one it was concluded that although there is growing evidence of the importance of biology to understanding the way alcohol and drug consumption is grounded in positive and negative reinforcement, and that certain characteristics related to this are inherited, existing research evidence does not support the conclusion that the social and cultural context of use is not also central to such use. Indeed Jellinek (1960) made clear the importance of cultural contexts to the development of problematic use, and there is little if any new evidence to suggest that this early view is incorrect. Plomin (1994) also makes clear the importance of environment, particularly with respect to the genetic mediation of psychological traits and environmental measures, but also with regard to direct genetic effects on environmental measures. In this context it is argued that there is an active relationship between genotype and environment such that the way people actively construct or select their environment is in part genetically determined. That is, the way people gain experience and subsequently select or construct their environment is in part inherited.

In terms of drug and alcohol use this has interesting consequences. Not only may certain people inherit a predisposition towards experiencing greater positive or negative responses to consumption, they may also inherit a predisposition to engaging socially with drug- or alcohol-using groups. Indeed, as already mentioned, Plomin (1994) found a high rate of heritability for substance use amongst peers from data derived from parental reports. Further corroborative evidence can be found from the findings of Tsuang *et al.* (1992) who, in a study of 1,626 pairs of twins from the Vietnam Era Twin Registry, found a stronger and more consistent rate of heritability for exposure to drugs than use of drugs. This study included use of or exposure to marijuana, stimulants, sedatives, cocaine, opiates and psychedelics (cited in Plomin 1994). If we also recall Crabbe and Goldman's (1992) conclusion that how a person reacts to exposure to alcohol problems in the home is partly responsible for determining alcohol-using outcomes, then the evidence for an active genetic component to alcohol and drug use, as opposed to a solely passive inherited predisposition to a particular risk-related form of metabolism, seems persuasive. Drug- or alcohol-related problems thus arise partly from an inherited metabolism which is extra-normative with respect to the way intoxicants are metabolised, especially in neurophysiological respects, and partly from the inherited selection or creation of specific environments. If we are to understand how some people encounter drug and alcohol problems then we need to understand by what means such selection and creation takes place.

All the evidence cited so far supports the conclusion that it is not all down to genes; the environment also plays a major role. However, the fact that environmental measures have genetic components cannot be bracketed out and the processes through which this takes place needs to be explored. We need a theory of how, in genetic terms, individuals' environments may be selected or created. To address this issue we need to review contemporary research on the evolution of the mind – that is, the way natural selection imparts certain cognitive characteristics. Darwin's theory of evolution lies at the centre of current thinking on the relationship between the environment and cognitive development; both in historical and contemporary terms. Hence, to enable the reader to most easily follow a review of work such as that of Edelman or Plotkin, it is first necessary to summarise the central principles of natural selection.

NATURAL SELECTION: DARWIN'S THEORY OF EVOLUTION

There are probably very few people alive today who have not heard of Darwin's theory of evolution. Many may not fully understand the central ideas behind the theory, and others may disagree with evolutionary concepts entirely, but the basic idea that contemporary biological species have been somehow transformed from more primitive ancestors is widely known in the Western world and elsewhere. The fact that evolutionary theory allows for the evolution of humankind from ape-like forebears and birds from dinosaurs is fairly commonplace knowledge in many parts of the world. Less understood is the way Darwin depicted this process; which is universally known as natural selection.

Unless one believes that there is an omnipotent force which constantly monitors changes as they take place on our planet and makes interventions designed to maintain viable life forms despite the generation of diverse environments, from ice ages to global warming, then one must accept that organisms adapt by some natural process to enable life to continue. Geological records are quite clear on the fact that the earth is very old, around 4.5 billion years, and continuously changing. Over time through changes in the pattern of Earth's orbit, dramatic climatic changes take place constantly, if rather slowly. At various times ice has covered most of the continent of Europe, whilst at others some areas have experienced sub-tropical weather. It is hypothesised from theories of plate tectonics that, in the distant past, all the continents of the world were joined together and slowly over time moved apart until some became separated by thousands of miles, as they now are. At the very least these changes would have produced climatic change, and, coupled with glacier formation and melting due to the occurrence of ice ages, it can be appreciated that environmental changes over time would have made strong survival demands on existing species. How biological species coped with such change is, in simple terms, the subject of evolutionary theory.

It could be argued that species survived the changes described above through migration. Like many animals on the plains of sub-Saharan Africa, species would migrate from areas sub-optimal to survival to more supportive ones. However, two important observations lead to this view being discounted. First, fossil records show that the more ancient a species is, the less it resembles contemporary species. From such records it has been possible to document the approximate time in the past of the appearance of distinct groups of animals. Some 2.5 billion years ago invertebrates started to appear; 370 million years ago fish emerged, from 225 million years ago reptiles, 30 million years mammals and, around 1.5 million years ago, the early ancestors of humankind are thought to have appeared. It may not require too much persuasion to convince the reader that what we see here is that over a very long period of time the diversity and complexity of species has increased. This being so, a theory which explains how this may have happened is needed, because if all the species which now exist and ever have existed were created together in some mythical past, then the fossil record would show a decrease in diversity and not the evident increase. It should be noted, however, that the earliest ideas about the fossil record in fact argued for a modified creationist theory – a theory in which periodic catastrophes wiped out all existing species followed by the re-creation of entirely new species by an omnipotent Creator. Eventually, the work of Charles Lyell (1797–1875) firmly established that geological evolution was not a discontinuous process, but was instead sequential and progressive. With this evidence becoming available from the 1700s onwards a climate within which ideas about evolution could be fostered formed. Although others such as Buffon (1707–78) and Erasmus Darwin (1731–1802), Charles Darwin's grandfather, developed ideas about the evolution of species, it was Lamarck (1774–1829) who produced the first explicit and comprehensive theory.

Essentially, the task Lamarck faced was to provide a theory which could account for the great diversity of species in existence, and which could also account for the fact that fossil evidence showed that new species came into being over time whilst others became extinct. It was probably generally appreciated at the time that offspring were a composite of two parents. Consequently it was possible to reason that in some way parents were able to pass on their characteristics to their children. However, if this were so then greater uniformity would have been evolved over time. Unless, that is, phenotypic changes which occurred in parents during their lifetime could also be inherited (recall that the role of genes in inheritance was not known at this time). In fact there was a school of thought which held that phenotypic changes which occurred during the lifetime of a parent could indeed be passed on through some blood factor to offspring. How would this work in practice? Well if, for example, two parent sharks developed larger tails fins through chasing a quickly moving prey then their offspring would (or might) be born with larger tail fins. This theory became to be known as the *inheritance of acquired characteristics.*

It is important to note that in formulating these ideas Lamarck argued that phenotypic change occurred as the result of environmental influences rather than a Creator. In terms of the emphasis on evolution through the interaction between an organism and the environment, Lamarck's views are consistent with those held today. His theory became to be classed as an instructionist approach to evolution in that adaptive characteristics arose in response to the environment; shark prey swim fast so sharks develop bigger tail fins. The environment 'instructs' certain changes in the organism to take place. This requires that the organism must have high endogenous potential for change. However, despite a great deal of research over a long period of time and many false claims, there has never been any reliable evidence found which supports the existence of this mechanism. So, despite the attractiveness of Lamarck's ideas in terms of relating evolutionary change to environmental factors, they do not provide a workable theory of evolution.

In 1859 Charles Darwin published one of the most important texts to have been produced during the development of the biological sciences. *The Origin of Species by Means of Natural Selection or the Preservation of Favoured Races in the Struggle for Life* provided an alternative to Lamarck which did not require environmental instruction and an extremely flexible endogenous core to account for diversity and change. As the title indicates, Darwin's ideas were based upon what was termed natural selection. This system of evolutionary change proposed in existence of three components of change:

1 overpopulation of the organism in terms of environmental support;
2 subsequent competition amongst offspring for survival;
3 inheritable variations in phenotype.

It is fairly evident that the first of these two requirements are widely present throughout both the plant and animal kingdoms. Putting aside for one moment the special case of extinction, it is generally the case that many more offspring are produced than are necessary for the survival of a specific species. For example, both frogs and salmon produce far more eggs than are required to maintain their respective populations. The vast majority are eaten by predators and some fail to compete successfully for available food. Only a few survive into adulthood. Similarly, plant species produce many times the replacement number of seeds for a given species. Some are used as a food source by various animal species including humankind, whilst others fail to thrive in competition for space with other plant species. In both these cases plant and animal offspring are subject to competition for scarce resources.

If all progeny were phenotypically identical, then which survived and which perished would be solely a matter of chance. However, given component three above, which allows for variations in phenotype between progeny, the likelihood is that some will be better suited to survival than others. Some offspring of an animal species may be faster than the norm in catching prey, or may be more resistant to a common disease, whereas some plant progeny may be better at obtaining water from the soil in near-drought conditions. Any of these phenotypic differences could impart a relative survival advantage in certain conditions. And if these phenotypic differences arose from differences in genotype then the advantages imparted by these characteristics would propagate throughout the species through the achievement of better survival rates. That is, those more suited to a specific environment by virtue of a phenotypic advantage which was grounded in a genetic change would have better rates of survival, and would thus have a better chance of surviving to produce offspring which would then also have better rates of survival. Over time the adapted variety would tend to increase in proportion to the non-adapted variety. In this respect the adaptive advantage of an individual can lead to the production of a variant species. That is, adaptive change at the level of an individual can lead to change at the population level. That is, phenotypic changes which impart survival advantage increase in frequency within the species: adaptation leads to speciation (the development of new species).

Of course Darwin was not able to explain how individual differences could be passed on to offspring. As discussed in chapter one, Mendel did not publish his work on inheritance until 1865. Indeed Darwin felt that the weakest component of his concept of natural selection lay in the absence of a mechanism whereby variation could be passed on to progeny. Nevertheless, while some objected to Darwin's ideas on religious grounds, his work on natural selection became widely accepted as being the basis of evolutionary change. And, over time, despite the growth of genetics and the development of molecular biology, this situation remains largely unchanged. In fact microbial genetics has probably provided the clearest evidence in support of Darwin's main principles of selection. A classic experiment used with undergraduate biology students to illustrate natural selection at work (at least my time as an undergraduate biologist) involves the bacterium *Escherichia coli* (*E. coli*) and an antibiotic. When a small number of *E. coli* taken from a single clone are cultured to increase bacterial density and then placed on a growth medium containing the antibiotic, all but a small number of bacteria will be destroyed almost immediately. However, through genetic mutation, a few will have developed resistance to the antibiotic and thus over a period of a few hours these resistant *E. coli* will rapidly increase in numbers, producing a new, streptomycin-resistant strain. Only very few resistant *E. coli* are required for a vast new population of resistant bacteria to be produced. This example illustrates that processes taking place at the individual level, in terms of the selection of a genotype which imparts a survival advantage, lead to the emergence of a new population. If there were other additional environmental factors acting differentially over time upon resistant and non-resistant strains, there would eventually form distinct new species. Hence, selection acting at the individual level can (ultimately) give rise to speciation.

One of the most important aspects of this perspective is the dynamic relationship which is depicted between genes and the environment. Organisms do not emerge independent of any external influences, but instead evolve in form in response to environmental conditions. The basis of this response is not the inheritance of acquired characteristics as Lamarck believed, but environmental, natural selection as Darwin argued.

EXPERIENCE AND NATURAL SELECTION

So far we have seen how alcohol and drug use is experienced via neurophysiological mechanisms which affect brain biochemistry. Alcohol and other intoxicant drugs act on primitive areas of the brain and change levels of neurotransmitters or otherwise affect neurological processes to induce certain pleasurable or unpleasant experiences. The extent to which differences in response to alcohol and other drugs are genetic has also been considered, and there is a great deal of evidence which suggests that certain differences in response are inherited. The question that remains concerns the extent to which drinking or drug-using outcomes for the individual depend on genetic as opposed to environmental factors. There appears to be some evidence from twin studies that those predisposed towards alcohol problems through the inheritance of family characteristics do not necessarily develop alcohol problems. From this and other evidence it is necessary to accept that both genetics and experience are important in determining outcomes.

In addressing this issue the work of Plomin and many others has been reviewed briefly and the findings presented overwhelmingly support the importance of both genetics and environment in determining behavioural outcomes. The evidence seems clear; a wide range of environmental measures and psychological traits have been found to have genetic components. Many things we do, from choosing friends to succeeding at school, are in some part inherited. Plomin (1994) even goes so far as to state that eventually specific genes associated with psychological traits and environmental measures will be identified. Indeed this is the current research agenda of many biological and medical drug and alcohol researchers. It is probably foolish to doubt that over the next ten or so years great progress will be made and that consequently Plomin's goal for research on genetics and behaviour will be met to an increasing extent. But even if this takes place, a question will remain regarding how genotype informs behavioural outcomes. That is, will it be suggested that genes set up part of the equation and chance takes care of the rest? Or will there be over-determined models whereby it will be argued that eventually our knowledge of human genetics will be so complete that every possible outcome will be explained by means of knowledge of our genotype? I suspect not, but posing these questions draws attention to the need that exists with respect to the generation of theory concerning the mechanisms whereby genetics and experience interact. Genetics will tell us what human biological form, and variations on that form, will arise from specific genes, but what we will also need to know is how this biological potential is acted upon by the environment to give rise to one outcome rather than another. As we have seen, even the best results from current research on genetics and experience show that most of the variance in psychological traits or environmental measures is experiential not genetic in origin.

Traditionally many social scientists have addressed these issues through ideas about intentionality, personality, the existence of free will or some other metaphysical property. It can often seem that it is believed by many that humankind has a biological template but beyond that there exists an independence of mind which transcends this material form and allows each individual to create through intention their own experiences and thus their own reality. Of course in some paradigms it is accepted that the external world can be constraining, but even then subjective experience, where it is acknowledged, can be written about as though it is some sort of disembodied property. What is needed, I would argue, is a science of the mind which accounts for the way consciousness is formed in the context of specific environments, but which also grounds mind in the biological mechanisms of the brain. Consciousness, the knowing engagement of the subject with the external world, is the key to the next stage in our task of uncovering the nature of problematic drug and alcohol use. The importance of this cannot be understated. As Searle has noted:

all of those great features that philosophers have thought of as special to the mind are similarly dependent on consciousness: subjectivity, intentionality, rationality, free will (if there is such a thing), and mental causation. More than anything else, it is the neglect of consciousness that accounts for so much barrenness and sterility in psychology, the philosophy of mind and cognitive science.

(Searle 1995: 227)

For Searle it is important to move away from inventing intrinsic characteristics of the mind which imply some transcendent metaphysical properties which exist other than in some extricable biological mechanism. Consciousness arises from knowable properties of the brain, even if these have not yet been uncovered. Searle states that:

In our skulls there is just the brain with all its intricacy, and consciousness with all its colour and variety. The brain produces the cognitive states that are occurring in you and me right now, and it has the capacity to produce many others which are not now occurring. But that is it. There are brute, blind neurophysiological processes and there is consciousness, but there is nothing else. If we are looking for phenomena that are intrinsically intentional but inaccessible in principle to consciousness, there is nothing there: no rule following, no mental information processing, no unconscious inferences, no mental models, no primal sketches, no 2.5-D images, no three-dimensional descriptions, no language of thought, and no universal grammar. In what follows I will argue that the entire cognitivist story that postulates all these inaccessible mental phenomena is based on a pre-Darwinian conception of the function of the brain.

(Searle 1995: 228)

The introduction of Darwin here is crucial, for what Searle is claiming is that the brain is like any other organ in the body in that it does not have properties which transcend its *evolved* physical form. That is, it is just like any other aspect of our bodies: the form which exists has arisen from natural selection. If this is so, then we have a means to explore further ideas about the mechanism through which genetics and environment give rise to experiential outcomes. And the indication here is that the environment acts in a selective manner during the formation of consciousness. Those actions or traits which have value in a specific environment will be selected over others which do not impart efficacious actions or traits. Essentially, it is being proposed that consciousness is a property of a biological organ, the brain, which has evolved a particular form as a result of natural selection, and that specific actions or conscious experiences which subsequently arise do so from further selection on inherent variation in mental function at an individual level during an individual's lifetime. Or to put it more simply, your brain has evolved to have considerable, but not unlimited, variation in a range of functions. Which of these develops for a given individual depends to a large degree upon which environment they grow up in. Specific functions, capacities, orientations and so on are inherited, but which emerge and which remain unexpressed is a matter of natural selection within the contemporary world.

This may seem a very bold and somewhat over-deterministic statement to make at this stage in the argument, but it is one which – I hope – the reader will be satisfied has been adequately argued by the end. However, there is still some way to go. To take the next step in this journey towards Chaos it is necessary to consider the work of Edelman and his theory of the evolution of the mind. This is the subject of the next chapter.

Chapter 3

The evolution of the mind

The idea of linking natural selection with human experience is not new. Since the early 1970s Gerald Edelman and colleagues have been working on a Darwinian perspective on the brain which has at its centre the role of natural selection in the genesis of mind. There have been references made to the work of Edelman earlier in this book and a number of promises made that at some later point his important work would be addressed directly. Now is the time to take on this difficult task. For the purposes of this current analysis it is necessary to show that notwithstanding the importance of biology and inherited characteristics to the development of alcohol and drug problems, nothing is wholly determined for an individual. This is meant in the sense that even if we knew everything about a person's genotype and their environment we still could not predict whether or not a specific individual will develop drug or alcohol problems. Prediction at the individual level is mostly guesswork. At the level of populations we can arrive at probabilities, but this is a property of groups and not individuals. It will take the whole of this book to demonstrate this, but the next step concerning the uniqueness of consciousness is a crucial milestone in the development of the argument. Edelman's work argues for a biological model of the brain based upon the Darwinian notion of natural selection; and in so doing he brings to the fore the importance of experience and value in forming basic neurological structures and thus cognition and consciousness.

For Edelman the brain is not a computer hardwired to process external stimuli, but is instead actively engaged in constructing the environment. That is, the mind is not a blank sheet upon which the external environment writes a program from which to actively engage the world, but instead we are born with certain capacities for action which serve to construct and interpret the world. The philosopher Immanuel Kant (1724–1804) was the first to point out the possible existence of what he termed a priori categories of the mind, categories which exist prior to experience and help order the universe of external stimuli, but Edelman takes the view beyond this and argues that the way mind engages the world is an evolved capacity subject to selection. And in this, the morphology of the brain, the evolved anatomy, is the fundamental basis of behaviour and consciousness. Edelman considered that natural selection was an elementary component of the genesis of mind and individual consciousness:

> The fundamental basis for all behaviour and for the emergence of mind is animal and species morphology (anatomy) and how it functions. Natural selection acts on individuals as they compete within and between species. From studying the paleontological record it follows that what we call mind emerged only at a particular time during evolution [thus]. . . . The centre of any connection between psychology and biology rests, of course, with the facts of evolution. It was Darwin who first recognized that natural selection had to account even for the emergence of human consciousness.
>
> (Edelman 1992: 41)

But how can this happen? By what mechanism can natural selection affect the way we think and see the world? To answer these questions we need to look closely at Edelman's arguments as they give one of the central keys to understanding how individual uniqueness may arise. First, the morphogenesis of the brain needs to be considered.

THE MORPHOGENESIS OF THE BRAIN

As the above quote from Edelman indicates, morphogenesis, the development of physical form, is central to his ideas about the evolution of the mind. It is contended by Edelman (1992) that an understanding of the evolutionary significance of brain anatomy is central to an appreciation of the role of natural selection in the formation of mind. To understand this argument it is necessary for the reader to be familiar with the process of morphogenesis – the way cellular development gives rise to certain anatomical forms in response to environmental influence.

Research on the development of embryos has shown that morphological development cannot be accounted for in terms of simple cell division. At certain stages in development cells begin to differentiate with regard to both type and function. So a theory which adequately describes this process of differentiation is needed. In general terms embryonic development takes place spatially. That is, cells with different characteristics are organised spatially to produce complex structures such as the eye or a hand. Understanding the way this happens, the mechanism whereby cells organise spatially to produce complex anatomical forms, must first be addressed.

So how does this take place? Well, not all the answers are available to us, but some of the basic principles have been described. As you will recall from chapter one, fertilisation gives rise to a single cell termed a zygote. After a very short period of time after fertilisation the single cell begins a period of cell division in a process called *cleavage* which gives rise to a hollow ball of cells known as *blastomeres*. The hollow ball of cells is known as a *blastula*. The cell division process continues beyond the formation of the blastula and begins to involve cell movement. The first stage of movement is termed *gastrulation* where cells located on the surface of the blastula migrate towards a narrow groove termed the *blastopore* where they travel over the lip and move inside. This process gives rise to the *gastrula* which has three distinct populations of cells upon which the ultimate adult anatomy founded. These populations are the outer *ectoderm*, the middle *mesoderm* and the inner *endoderm*. The ectoderm eventually forms epidermal and nervous tissue via embryonic induction, a process whereby the eventual fate of embryonic cells is fixed. Nervous tissue first appears as a flattened plate of cells known as the *neural plate*, the cells of which fuse to form a *neural tube* which ultimately becomes the central nervous system. Throughout this process cells divide, move and differentiate. A complex process which requires careful orchestration to produce a functional adult member of the population. Edelman has summarised the important characteristics of this process:

1 Cells *divide*, passing on the same amount and kind of DNA to their daughter cells.
2 Cells *migrate*, separating from their connections in sheets called epithelia to form a loose, moving collection called a mesenchyme. (The sheets themselves can also move by curling

up into tubes without releasing the contacts between their cells).

3 Cells *die* in particular locations.

4 Cells *adhere* to each other . . . or they lose their adhesion and migrate to another place. This migration occurs on the surface of cells to form layers, or on matrix molecules released by the cells. The cells then readhere, forming new combinations.

5 Cells *differentiate*; they express different combinations of the genes present in their nuclei. They can do this at any time and place but only if they receive the right cues. Only certain places in the developing embryo have the right cues. This process of differential gene expression is called differentiation. It is what makes liver cells different from skin cells, and skin cells different from brain cells, and so on. Differentiation means specific patterns of protein production; some genes specifying particular proteins are turned on and some are turned off. Each cell of a given type has many proteins, only some of which are shared with cells of a different type.

(Edelman 1992: 58; italics in the original)

This summary draws attention to the complex interrelationship between different adjacent cells. Spatial orientation is important in that it is the position of one cell in relation to those around it which determines the developmental fate of a cell. That is, whether a group of cells will effectively become a foot or a kidney. Embryonic induction thus has both temporal and spatial components. In consequence of the specific timing of movement of cells and their adhesion to others, particular shapes are formed. As genes which give rise to what are termed *morphoregulatory* molecules are activated and deactivated in set sequences, cells move, adhere, fold and/or are prevented from further movement such that specific shapes arise. Given that different species have different genotypes then different shapes will arise for different species; not only in terms of overall appearance, but with regard to the smallest components of morphology. This emphasis on spatial orientation led Edelman to term this branch of developmental studies *Topobiology*.

You may at this stage be once again asking yourself why you are being presented with all this detail. Well, the issue is one of complexity. If the morphology of an adult human being is dependent upon the intricate processes described above than it can reasonably be expected that they will not unfold in exactly the same way for each individual, not even perhaps for identical twins who share all the same genetic code. Any small difference in the positioning or folding of a cell or group of cells would result in a different cellular environment for adjacent cells. Although there are species-level limitations on the viability of morphological change, individual changes will nevertheless take place. As Edelman has so clearly stated:

> Notice the main features of this drama. It is topobiological, or place-dependent. Events occurring in one place require that previous events have occurred at other places. But it is also inherently dynamic, plastic, or variable at the level of its fundamental units, the cells. Even in genetically identical twins, the exact same pattern of nerve cells is not found at the same place and time. Yet the collective picture is species specific because the *overall* constraints acting on the genes are characteristic of that species.
>
> (Edelman 1992: 64; italics in the original)

This idea of diversity at the individual level is central to Edelman's theory of the mind. As was discussed above, biological diversity is acted upon by the environment by way of selection in that some chance forms give rise to survival advantages. Put in the most simple terms the keys are variation and competition. What works best is selected whilst less effective variations do not achieve as high a level of survival. This is the principle of natural selection which, Edelman argues, applies to the development of the mind through the topobiological morphogenesis of the brain. This is a complex issue which requires some careful consideration.

NATURAL SELECTION AND THE MIND

The first component of Edelman's ideas consists, as we have seen, of a theory of the morphogenesis of the brain which argues that topographical features, such as place and timing, are central. In consequence of this it is further stated that the actual configuration of micro-components of adult anatomy, such as the fine detail of neural networks, cannot be known. Although there are overall species constraints placed upon what is possible, there is nevertheless potential for differences to occur at the level of the individual. This anatomical diversity is held to be evidence that the brain cannot be a mere computer-type system which follows external instructions. Such a system would require considerable conformity of form to enable it to meld efficaciously with the external world. Recall that natural selection holds that evolved forms impart relative survival advantage, a system that was hardwired to one specific external world would not do so. Even minor changes could effect the extinction of such organisms. In contrast, a system of brain morphogenesis which imparted diversity in brain anatomy, and thus mode of function, would impart survival advantages. Accordingly, Edelman argues that his topobiological account of brain development, based as it is on a selective rather that instructionist basis, corresponds more closely with what we know about biological morphogenesis, natural selection and inter-species anatomical diversity. If this perspective is not persuasive enough, Edelman proposes a further reason for adopting the selectionist outlook:

> A potent additional reason for adopting a selective rather than an instructive viewpoint has to do with the homunculus. . . . the little man that one must postulate 'at the top of the mind', acting as an interpreter of signals and symbols in any instructive theory of the mind. If information from the world is processed by rules in a computer-like brain, his existence seems to be obliged. But then another homunculus is required in *his* head and so on, in an infinite regress. Selectional systems, in which matching occurs *ex post facto* on an *already existing* diverse repertoire, need no special creations, no homunculi, and no such regress.
>
> (Edelman 1992: 82; italics in the original)

Thus Edelman does not call upon matching analogies or appeal to mystical and regressive components of the brain. He instead argues for a theory of mind which can be seen to reside in a brain structure and function which is firmly anchored in existing biological knowledge of developmental processes. The theory he has developed in line with this somewhat formidable requirement he has termed the theory of neuronal group selection (TNGS). This is a complex theory, but fortunately its basic canons are only three in number. These are:

1 *Developmental selection.* The processes of morphogenesis discussed above, which depend upon cell division, movement and

differentiation, give rise to a species-dependent neuroanatomy which is nevertheless variable at the individual level. Cell lines which survive to form specific anatomical features such as neurons do so at the expense of others, in this sense the process is selectional in that groups of cells are involved in what is termed 'topobiological competition'. Groups of neurons in a specific brain region which arise from this selection process are known as *primary repertoires*. Hence developmental selection holds that the genetic code provides a template which imposes constraints upon anatomy but does not impart anatomical uniformity. Neuronal network diversity is the result of this process.

2 *Synaptic selection*, the second component of TNGS, is based upon the promotion or neglect of synaptic connections. Through behavioural engagement with the external world some synaptic connections are promoted through use and are thus strengthened, others are neglected and are thus weakened. By these means certain neural networks are selected from the diversity of those available. These selected networks or circuits are called *secondary repertoires*. Edelman argues that in practice primary and secondary repertoires may overlap developmentally.

3 The third and final component is perhaps the most complex. In consequence probably the best way to proceed is to quote directly from Edelman and then offer my own somewhat simplified account of what the reader needs to think about in terms of the more modest analysis presented in this present book. Edelman (1992) states that the final tenet of his theory is concerned with *how primary and secondary repertoires link psychology and physiology*. The concern is with the means by which fundamental properties of brain development give rise to a structure in which known human psychological properties come into being. That is, how do evolved brain areas coordinate to produce the functions of the mind? For Edelman, for such functions to be carried out:

> primary and secondary repertoires must form maps. These maps are connected by massively parallel and reciprocal connections. The visual system of the monkey, for example, has over thirty different maps, each with a certain degree of functional segregation (for orientation, color, movement, and so forth), and linked to the others by parallel and reciprocal connections. Reentrant signalling occurs along these connections. This means that, as groups of neurons are selected in a map, other

groups in reentrantly connected but different maps may also be
selected at the same time. Correlation and coordination of
such selection events are achieved by reentrant signalling and
by the strengthening of interconnections between the maps
within a segment of time. A fundamental premise of the TNGS
is that the selective coordination of the complex patterns of interconnection between neuronal groups by reentry is the basis of
behavior. Indeed, reentry (combined with memory . . .) is the
main basis for a bridge between physiology and psychology.

(Edelman 1992: 85)

So how is this linked together to give a theory of the mind? Well, at
the basic level Edelman is arguing that environmental factors act at
multiple levels to select some configurations of neural cells over
others. At the level of primary repertoires it is a stochastic process
in that cell lines, in competition with others, either do or do not survive. Within secondary repertoires, selection by the environment of
specific synaptic pathways occurs through behavioural engagement
with the outside world. And the third component describes how
primary and secondary repertoires combine to create maps, and
how different maps are connected via synaptic cross-linkages.
What Edelman considers takes place is that through selection at
the level of both primary and secondary repertoires neural maps
are formed which underlie a particular brain function. Now these
maps are not held to be distinct and therefore autonomous, but
are instead linked routinely through widespread proliferation of
neural fibres in the developing brain tissue. Many linkages that
are formed in this way will ultimately serve no purpose, but
others will be selected for because they will impart some survival
advantage to the individual. For example, imagine for one
moment that there is a neural map the function of which is the perception of liquids, and there is another map with a motor coordination function around hand movement. Now if these two maps are
unlinked then, in terms of this illustration, we could say that the
individual would be able to see the liquid and move their hand
but without some third factor which coordinates these behaviours
it would not be possible to move the hand to get the water. There
needs to be either a neural connection between these two maps or
there needs to be some kind of homunculus that can link one
thing with the other. Now Edelman argues that it is neural linkages
which enter and reenter connected maps which give rise to coordi-

nated action such as that described. There is no need for a third higher coordinating function. In considering even more complex action it is possible to imagine greater and greater numbers of re-entry connections between vast numbers of neural maps formed through the selective development of primary and secondary repertoires. Edelman describes this as a dynamic process whereby new signals from the outside world can be reentered into existing maps as experience of the external world grows. Edelman has simulated this process in a computer model (the reentrant cortical integration model) which corrclates many different maps through reentry, and he has found that such a model can produce coordinated responses to complex visual presentations (see Edelman 1989).

The reentered connectiveness of individual maps is the first order of mapping. The way large groups of maps are attuned is through what is termed by Edelman a *global mapping*. A global mapping is a dynamic structure containing many sensory and motor maps which can interact with unmapped areas of the brain such as the brain stem. These global maps serve to allow: 'selectional events occurring in its *local* maps . . . to be connected to the animal's motor behaviour, to new sensory samplings of the world, and to further successive reentry events' (1992: 89). Hence a global mapping is a higher-order structure the function of which serves to categorise cognitively the external world in reference to evolved functions of the brain stem and other evolutionarily more primitive areas of the brain. Now, as was discussed in chapter one, these lower areas of the brain are involved in a variety of forms of experience such as emotive arousal or pain, so it would seem from the above that a global mapping has incorporated within it a *value* component. That is, the selected sensory/motor function of reentered maps is incorporated through a global mapping with value; something which is good or bad (or perhaps indifferent). Edelman is quite clear on this point:

> sensorimotor activity over the whole mapping *selects* neuronal groups that give the appropriate output or behavior. . . . But what is 'appropriate' with respect to behavior, and how does perceptual categorization manifest itself? The TNGS proposes that categorization always occurs in reference to internal criteria of value and that this reference defines its appropriateness. Such value criteria do not determine specific categorizations but they constrain the domains in which they occur. According to the

theory, the bases for value systems in the animals of a given species are already set by evolutionary selection. They are exhibited in those regions of the brain concerned with regulating bodily functions: heartbeat, sexual responses, feeding responses, endocrine functions, autonomic responses. Categorization manifests itself in behavior that appropriately fulfils the evolutionarily selected requirements of such life-supporting physiological systems.

(Edelman 1989: 90; italics in the original)

He makes this point even more unequivocally when he states that: 'Accordingly, the TNGS, the driving forces of animal behavior are thus evolutionarily selected value patterns that help the brain and the body maintain the conditions necessary to continue life' (p. 94). Thus it is being argued that the way we engage the world perceptually and behaviourally is selected for in terms of survival advantage. An action which has value will be selected for over one which does not. This is a complex idea, but it is one which may become clearer through use of an imaginary example.

Human beings need to drink water. Without an appropriate daily intake of water a person will eventually experience impaired function and ultimately die. In consequence it is not unreasonable to suppose that during the evolution of land-based organisms the consumption of water became a selected trait or behaviour. Those early variants of fish which did not develop this action did not evolve into land-based species. Now, as we know from some of what was said in chapter one, in general it is much easier to get people to do things which are pleasurable than either unpleasant or neutral in this regard. It is easier to imagine getting people to self-administer morphine over a long period of time than, say, an antibiotic. Hence, if water tasted unpleasant it might have been the case that land-based species never would have evolved because their water intake would have been too low to sustain life. But, as we know, when thirsty, water tastes extremely pleasant, hence it is meaningful to describe it as an activity with positive value. It is possible to assert with some confidence that this has probably been the case for mammalian species for millions of years. In this sense, then, the value of water is an inherited trait which has been selected for during some time in the evolution of the human species. The consumption of water is, broadly speaking, what Edelman would term a value criterion.

This example refers to an historical case of evolution through natural selection, whereas Edelman is primarily concerned with contemporary natural selection in the formation of perceptual and cognitive function. That is, the selection of global maps. So how does he propose this takes place in practice?

Staying with the example of water, we need to visualise a new-born (or at least very young) child developing the ability to collect water in a hand and moving this to its mouth to drink. Well, it is argued from Edelman's perspective that initially the child will move its hand around at random through the excitation of many synaptic pathways associated with movement of the hand. Eventually by chance the hand will touch the water, and given that water has value this action will have preference over those actions which do not result in the water being touched. As this action repeats over time those synaptic pathways which inform this movement will be selected for, and thus reinforced in function, by virtue of their inherent value over those which do not. These pathways will be reentrant pathways between visual and motor maps which are globally mapped to those parts of the brain concerned with thirst. This process would continue stochastically until the hand reaches the mouth and the water is consumed. This global mapping of water consumption would in effect arise from natural selection; the unit of selection being a synaptic pathway.

This is an extremely powerful analysis which has been demonstrated to work in practice through computer modelling. Edelman provides an example:

A specific example of categorization constrained by value may help connect these ideas. My colleagues and I have simulated complex automata based upon the TNGS in supercomputers to demonstrate that perceptual categorization can be carried out on value in global mapping. . . . In automata such as Darwin III, value is seen to operate for the visual system, for example, in circuits that favour light falling on the central part of the eye. (Value = 'light is better than no light'; light and stimulation at the center of vision are favoured over light and stimulation at the periphery.) In Darwin III, the action of these value circuits enhances the probability that synapses active when such circuits are engaged will be strengthened in preference to competing synapses. The net result is that with selection and experience the eye of the automaton tracks signals from lit objects. This defines one form of 'appropriate' behavior as acquired behavior that is consistent with evolutionarily set values.

(Edelman 1992: 91)

Edelman (1992) is quite clear on this interrelationship between value, natural selection and perceptual categorisation.

Edelman argues that the basic principles of the TNGS as outlined here lie at the foundations of all perceptual categorisation, memory and learning. It is not necessary here to go into the fine detail of the areas of concern, but instead simply to note the importance of the integration of different brain areas in these cognitive events. Memory arises from alterations in the synaptic strength of neural groups in a global mapping. That is, the basis of memory is neuro-physiological changes which occur in response to experience of the external world. Through the interrelatedness of neural maps new experiences can trigger old maps; crudely, this may be one key basis of memory. It is important to note, however, that this is a dynamic process as the complex interconnectiveness of neural maps and global mapping can mean memories themselves and patterns of association can vary over time. We might imagine from this that learning merely arose from perceptual categorisation (say, identifying water) and memory, but according to Edelman what is needed are further value centres, involved in 'expectancy'. By expectancy Edelman means the lack of satisfaction of the 'set points' of physiological structures to be found in the hedonic system (e.g. the hypothalamus, limbic or hedonic system, midbrain regions of the evolutionarily more primitive areas of the brain). By set points Edelman means evolved levels of response from these centres (selected by virtue of increased fitness). Learning in this context is about linking categorisations to behaviours which impart a survival, adaptive advantage. In other words, learning is about remembering which matched pairs or groups of behaviours and perceptual categorisations satisfy evolved values. Then, as memory is dynamic and adaptive, so too is learning. Indeed, greater brain complexity in terms of increased numbers of neural maps, interconnectiveness and global mapping is, therefore, also adaptive.

Edelman argues that through the increased complexity over time which arose from further stages of selection, these basic processes enabled animals to locate and recall events and locations which promoted their survival. But in human beings further developments took place which gave rise to consciousness. The theory proposes that the two major organisational units of the nervous system, the limbic-brain stem and the thalamocortical systems were linked during evolution and the consequent improved capacity to respond to complex environments imparted a survival advantage. In this way consciousness is held to have arisen through the same process of natural selection that produced more primitive responses to, say, light or warmth. As the reader will be aware from chapter one, the brain stem and limbic systems of the brain are involved with hunger, sex, sleep, pain and other similar functions. The thalamo-cortical system is involved with the rapid and multi-layered processing of sensory information, such as sight, smell, hearing and motor awareness (where parts of your body are in relation to the outside world). Edelman holds that these two systems were linked during evolution such that:

> The later-evolving cortical system served learning behavior that was adaptive to increasingly complex environments. Because this behavior was clearly selected to serve the physiological needs and values mediated by the earlier limbic-brain system, the two systems had to be connected in such a way that their activities could be matched [such matching would impart an advantage in survival terms hence it would have been subject to selection]. Indeed, such matching is a critical part of learning [which imparts advantage]. If the cortex is concerned with the categorization of the world and the limbic-brain stem system is concerned with value (or with setting its adjustments to evolutionarily selected physiological patterns), then learning may be seen as the means by which categorization occurs on a background of value to result in adaptive changes in behavior that satisfy value.
>
> (Edelman 1992: 118)

These processes also lie at the root of the emergence of what Edelman terms primary consciousness. Through the mutual interaction of the two systems, the linking of the brain stem and limbic systems with the thalamocortical systems gave rise to the possibility of a higher-order perceptual response. This is termed 'value category' memory in that recall of a perceptual categorisation occurs within the context of an evolved value. These memories are forged from the merging of primal values with sensory perception. This gives rise to the successful engagement with the world through past valued knowing; that is, value memory. This in itself does not result in primary consciousness. However, the reentrant mapping of this value memory with the real-time engagement with the external world does give rise to a form of knowing termed by Edelman the 'remembered present':

> The brain carries out a process of conceptual 'self categorization' [self-referring internal systems]. Self categories are built by matching past conceptual categories with signals from value systems; a process carried out by the cortical systems capable of conceptual functions. This value-category system then interacts via reentrant connections with brain areas carrying out ongoing perceptual categorizations of world events and signals. Perceptual (phenomenal) experience arises from the correlation by a conceptual memory of a set of ongoing perceptual categorizations [the internal systems – self – becomes 'aware' of the external passage of events]. Primary consciousness is a kind of 'remembered present'.
>
> (Edelman 1992: 119)

In summary, perceptual categorisation and the resultant coordination of function occurred through reentrant mapping between maps of neural cells and synaptic pathways. Global maps emerged, and these served to ground such perceptual categories in an evolved system of value. The evolution of the thalamocortical system concerned with the multi-layered processing of sensory information provided the next step, followed by the interconnection of this system with the more primitive brain systems concerned with value. The mutual interactions of these two systems gave rise to a special form of memory termed 'value memory'. Through the reentrant connecting of this memory system with the continuous sensory monitoring of the outside world a primary consciousness arose as a 'remembered present'. The driving force of these changes was natural selection. In consequence, participation in the world became more efficacious.

Basically, once the ball started rolling with the evolution of a distinct nervous system structured through neural networks and based on synaptic pathways, the ever-increasing complexity we see from reptiles through to human beings can be understood as a process of selection on elaborations on a basic theme. This imparted (and still imparts at the level of synaptic selection) greater coordination between systems concerned with physiological value and those which function to coordinate the organism with the outside world.

So, we are nearly at the end of this brief exploration of the central themes of Edelman's work. All that is required now is to review his theory of the emergence of higher-order consciousness in terms of self-awareness and language. This is not an easy area but careful attention to detail may lead to further cautious progress. This long diversion from alcohol and drug use may seem to some to be somewhat over-extended, but I am confident that patience will bring just rewards.

With respect to the first level of cognitive development, the reader has already been introduced to the concept of *perceptual categorisation*: an unconscious process of interconnecting maps of perceptual events. The second level of development involves the evolution of global mapping which connects perceptual categorisation with memory (the accessing of preexisting neural maps) giving rise to *conceptual categorisation*. Third, by linking this with each sensory modality, *primary consciousness* came into being. Think of this as being, through memories of this special kind, a form of recognition of present sensory inputs. Also recall that through linkages with evolutionarily more primitive areas of the brain, primary consciousness arose within the context of species-related value. Note that this account of the evolution of the brain states that the capacity for conceptual processing exists prior to the development of lan-guage – a conclusion I think Immanuel Kant would have warmly welcomed!

The development of language is, however, seen by Edelman to be indispensable to the development of higher-order consciousness. Edelman argues that through the action of selective events, evolved phonological capabilities were being linked through learning with concepts and non-verbal behaviour. From this developed a vocabulary of sounds which had meaning in the sense that they depicted external events which had intrinsic species-related value. (I may have gone beyond Edelman's exact composition of these events, but I think it likely that the emergence of value-sounds would be a first step towards the formulation of words. There may be greater clarity in thinking of words as a higher-order property. The 'aaaah' we emit when we fall may be a value sound, but it certainly is not a word in the usual sense.) According to Edelman a syntax then arose: 'by connecting preexisting conceptual learning to lexical (vocabulary) learning' (1992: 129). This is termed 'semantic boot-strapping' to illustrate that this happens within and from itself without any universal grammar or hardwired programming.

In some ways I find this one of the least satisfying aspects of Edelman's theory because it seems to suggest that something is created out of nothing. The problem stems from the fact that he has previously argued that concepts and gestures were linked with phonological capabilities through learning to form semantics. But later it is contended that through being connected, conceptual learning and the lexicon give rise to syntax. Surely, if phonological capabilities, concepts and behaviour give rise to value sounds (or as Edelman would claim, words) which form a lexicon or vocabulary, then something in addition to these elements must be added to the equation for a new state to emerge; that is, for the development of syntax to take place. Perhaps the clue is in the use of the term 'pre-existing', for it would seem reasonable to propose that reentrant connectiveness between existing global maps and neural maps involved with the recall of a value sound or 'word-situation' (an additional special form of memory) would give rise naturally to developments of the form of 'word-word-situation', 'word-word-situation-situation', 'word-situation-word-situation' accumulations. Such accumulations would of course be a primitive syntax. Edelman even identifies the brain areas which may give rise to this further particular form of memory: Broca's and Wernicke's areas. If this is so then the term 'bootstrapping' is misleading as no new processes other than reentrant connections are needed to support the proposal. Edelman later states that:

> Because of the special memory provided by Broca's and Wernicke's areas, the phonological, semantic, and syntactical levels interact directly and also indirectly via reentrant circuits that are formed between these speech areas and those brain areas that subserve value-category memory. When a sufficiently large lexicon is collected, the conceptual areas of the brain categorize the *order* of speech elements, an order that is then stabilized in memory as syntax.
>
> (Edelman 1992: 130, italics in the original)

However, there still seems to be an unacknowledged plea for the existence of previously undiscussed cognitive capabilities. (How do the conceptual areas of the brain categorise?) It should be noted, however, that the problem may be one of my own making rather than being present in Edelman's work! Nevertheless, whatever the exact mechanism, Edelman is correct to conclude that the acquisition of syntax heralds the emergence of symbolic memory.

The emergence of a symbolic memory takes us almost to the end of this journey through Edelman's ideas. From the interaction between extra-linguistic conceptualisation, value memory, emergent symbolic capability and a social speech community a concept of 'self' can arise. Given that the brain structures from which these functions arise are infinitely variable between different individuals by virtue of the selective conditions of morphological development, the 'self' is always unique even in highly ordered environments. A point Edelman has made clearly:

> While the embodiment of meaning and reference can be related to real objects and events by reentrant connections between value-category memory and perception (primary consciousness), simultaneous interactions can also occur between a symbolic memory and the same conceptual centers. An inner life, based on the emergence of language in a speech community, becomes possible. This is tied to perceptual and conceptual structure, but it is highly individual (indeed, it is personal) and it is also strongly tied to affect and reward. It is higher-order consciousness, capable of modelling the past, present, future, a self, and a world.
>
> (Edelman 1992: 133)

The importance of these conclusions cannot be overstated for they are based upon a clearly reasoned explanation of consciousness and self-awareness as being the result of evolutionary forces arrayed within an inter-species consociate order. If human beings were a solitary species like many other mammals, then language could not have emerged, and hence nor would have conscious awareness. Only by way of a social, collective species could the emergence of phonological abilities have led to language and a symbolic memory.

Instead of being a mere computer program or an epiphenomenon of mental computation, consciousness is seen within this perspective to be a product of natural selection acting upon brain morphology. Each stage of development from simple neural maps through to syntactic memory imparted a survival advantage upon human beings as a collective mammalian species occupying a changeable and unpredictable world. There are other possible solutions to this problem. For example many insects, such as the cockroach, benefit in survival terms from high rates of reproduction which give rise to greater resources in terms of species level genetic variation. In consequence there is an increased probability that a change in the environment can be met by a suitable variant genotype. It is by this relatively more simple mechanism that bacteria develop resistance to antibiotics and mosquitoes to insecticides. By comparison, the evolution of intelligence is a more complex and precarious path, but nevertheless the adaptive basis of both seemingly different events is the same.

The second important element to the argument is that despite the fact that the structures of the brain which give rise to conceptual abilities and self-consciousness arise at the level of species, there is no uniformity between individuals. The morphological development of the brain is a stochastic process such that the exact form of specific synaptic pathways which underlie neurophysiological structuring varies between individuals, even in the case of identical twins. In consequence it can be asserted with confidence that no two individuals will experience the external world in exactly the same way. Edelman (1992) draws this point out clearly when he claims that the collection of individual and subjective experiences, feelings and sensations associated with awareness are unique to each individual. These experiences, feelings and sensations are termed *qualia* and they will be subtly different between individuals because the same synaptic pathways which underpin neural structures are never repeated between individuals. Differences in primary and secondary repertoires will lead to differences in the formation of maps upon which perceptual categorisation is based. If maps are different then so too will be global maps, and, given that these project to value centres of the brain, it can therefore be proposed that value too is a variant property. Moving onwards to consider the structural basis of higher-order consciousness and self-awareness, it can be assumed that these same principles of embedded variance arise such that diversification between individuals can be expected. Of course it needs to be remembered that the variation that occurs is always within species-level constraints, as discussed above. Departures too far from the norm would be likely to be non-viable.

TOWARDS SOCIETY

So far this has been a long a complex journey through a considerable amount of material relating to the biology and genetics of alcohol and substance use, the relationship between genetics and experience and, latterly, Edelman's theories of the evolution of the mind. No doubt many readers are beginning to wonder when the discussion is going to return to the main theme of alcohol and substance use; that is, at what point is this analysis going to relate directly to the subject in hand? I would like to remind you of the original construction of the problem given in chapter one. The question posed was why do some people use alcohol or drugs more than others? Well, I think we are getting somewhere even though it has not yet been made explicit. From chapter one we know that different people have different neurophysiological responses to alcohol and other drugs, and that part of such differences can be inherited. From chapter two it can be appreciated that although differences in response to alcohol and other drugs are genetic in origin this is only part of the story. Work on the relationship between genetics and experience has demonstrated that genetic differences only account for a minority of the variance in environmental measures. It would seem that even genetically identical individuals experience different outcomes. Now how can this be so? Plomin (1994) draws our attention to the fact that, at least in part, people create their own environments and are not merely passive recipients of external forces. As Plotkin (1994) argues so well, intelligence generates the causes of its own behaviour: 'Once intelligence has evolved in a species, then thereafter brains have a causal force equal to that of genes' (p. 177). And as we have learned from Edelman, this intelligence differs between individuals. We have a series of clear findings here, all of which point to the existence of diversity in biology, brain, mind, cognition, self-consciousness and behaviour. But so far little has been said about the social world. What we need now is to unfold this analysis further in the context of collective social action in the contemporary world. Some understanding of the mechanisms whereby people differ from each other in important ways has been gained. It is time to situate this in social and cultural outcomes.

Chapter 4

Consciousness and language

In chapter four, as in all the previous chapters, there is a strong focus on uncovering important aspects of human biology as they relate to alcohol and drug use. True, the connection has not always been obvious in places, but nevertheless important progress has been made in grounding human action in aspects of biological form. The embodiment of human action is an important step in moving towards a comprehensive theory of human existence. Too often natural scientists conduct their research as though humankind were no different from non-conscious animals, and social scientists too frequently present their work as though we were some disembodied consciousness unrestricted by our body and its constraints. I make no complaint about the way either natural or social scientists pursue their quests for knowledge. Good empirical research practice does not require that the investigator be constantly aware of universal perspectives. However, at the level of theorising it is essential to recall that humankind is an evolved biological, collective species within which there has emerged a higher-order, symbolic process of conceptualisation which we experience as conscious awareness. Everything about us should ultimately be accounted for in terms of all these elements. We are some way towards our goal, but there is still a long way to go.

In terms of this present text we have spent long enough within the biological domain, so it is time now to consider the more social contexts of the human condition.

THE GENESIS OF THE COLLECTIVE SELF

Despite the complexity of Edelman's ideas, one thing will have drawn the readers attention; that the 'self' arises from social interactions based upon language. Now little of the broad biological content of his work may have been familiar, but the idea that language forms the self is certainly not new as a general proposition. Indeed both Wittgenstein and, much later, Lacan have language at the centre of their theories of mind and being. In this regard, Edelman can be seen to have provided a grounded context for much earlier philosophical and psychological work.

According to Schatzki (1993) one of the key components of Wittgenstein's perspective on the mind and society centres on the notion of 'conditions of life'. These are held by Schatzki to be: 'an aspect or component of how things stand or are going on in a person's life. The term "condition" is used here in the sense of the state of something's being, the "how-it-is" of something' (p. 285). It is later argued (p. 295) that for Wittgenstein these conditions of life are causally grounded in neurophysiological structure. In this case, then, perhaps it would be less obscure if 'conditions of life' were less mystically referred to as species-dependent mental processes depicted by Edelman as perceptual categorisation, value memory and primary consciousness. If human beings can be reduced to a notion of 'how they are' then these evolved elements of the mind must be a primary source of ontological status. In fact this non-social component of mental life appears to have been openly recognised by Wittgenstein in some form through his articulation of the existence of inner phenomena which are not accessible to an external observer. In contrasting inner phenomena with external events Schatzki states that:

> The categorical difference between inner episodes and objects in the world [stems from] the fact that inner episodes cannot be observed in the way objects can, namely, by attentively following alterations in their features. Inner episodes are observed only in the sense that they occur consciously. . . . A less metaphorical way of putting this point . . . is that inner episodes are the appearances of a person's conditions to himself.
>
> (Schatzki 1993: 292)

Now this insight is very similar to Edelman's conception of 'qualia'. Edelman (1992) remarks that: 'Qualia constitute the collection of personal or subjective experiences, feelings, and sensations that accompany awareness. They are phenomenal states – "how things seem to us" as human beings. For example, the "redness" of a red object is a quale' (p. 114). As we would expect from the unique way each brain is formed morphologically, qualia are individual. No two people can be expected to see the same thing in exactly the same way: 'What is directly experienced as qualia by one individual cannot be fully shared by another individual as an observer' (p. 114).

But Wittgenstein also described outer mental episodes. The fact of mind, that it exists, may be the consequence of biological existence, but what actually happens to and within a person is an outcome of social engagement: 'Mind is a social institution, the body that expresses it is socially moulded, and the interests that guide these processes concern sociality' (Schatzki 1993: 309). The use of the term 'moulded' is important as it depicts an interaction between the social world and the mind, and not a simple determinism whereby the mind is somehow a blank sheet upon which society writes a script. Consciousness, being based upon individual qualia, or inner episodes in Wittgensteinian terms, is always unique to a particular individual but is also, through the acquisition of language, partly shared as well. Higher-order consciousness emerges within a language community but it is grounded elsewhere in particular biological structures. The former comes into existence through the circumstances of the latter, as Schatzki disclosed:

> Physiology and biology only undergird, as a matter of both historical and continuing fact, the possibility of expressive bodies manifesting mind. They accomplish this by housing the physical causality responsible for all bodily behaviours and sensations (and not only naturally expressive ones). But this does not reduce mind to these facts or bodies. Mind is a social institution carried away in discursive practices. Our physical being causally brings about the activities comprising these practices and thereby makes mind possible.
>
> (Schatzki 1993: 302)

It is hard to imagine a clearer philosophical statement of Edelman's position. There would appear to be little difference with regard to these conditions of mind between Wittgenstein and Edelman except with regard to the language used. Change the term 'mind' as used here by Schatzki to 'higher-order consciousness' and the similarities become even more evident.

For both Wittgenstein and Edelman the human condition is structured through internal cognitive processes involving distinct, individual, perceptions of the external world. The individual component of experience, termed qualia (Edelman) or inner episodes (Wittgenstein), arises causally from the neurophysiological structures of the brain. Outer phenomena, or higher-order consciousness, are also grounded biologically, but they are given specific form, or moulded, by social events and institutions. What we are for ourselves, what the self is, is in part an evolved way of perceiving and valuing the world, and in part the consequence of the ability to represent the world symbolically through the acquisition of a shared language. Mind has both unique and shared properties: unique in that the brain achieves form by way of stochastic and selectional processes which give rise to different qualia and thus different ways of perceiving the external world; shared in that internal value arises evolutionarily through natural selection, and from the fact that self-consciousness, though grounded biologically and globally mapped to value memory and perceptual categorisations (which give rise to distinct qualia), arises with a language community.

There thus seems to be strong agreement between the positions of these two thinkers from very different traditions. This is significant as in terms of this present analysis it is important to establish clearly that human existence *in terms of the way life is actually lived*, the things we do and do not do, though determined through clearly perceived biological processes is nevertheless uncertain in outcome. It has been disclosed that mental life is at the same time both given and unknowable; it is a complex phenomenon which is both biological and social. But now is the time to move on a step and try to say something about what kind of knowing these events in our morphogenesis as thinking, conscious and intentional beings has given rise to – in other words, it is necessary to move further into the realms of the social world and describe what we can expect from the above in terms of the nature of our individual being situated as it is within a social, collective symbolic order.

INSTINCT AND REASON

The central idea expressed above with regard to the emergence of a consciously aware selfhood is that the means for this to take place are grounded in the existence of a collective symbolic order jointly experienced through a shared language. As Wittgenstein (1973) claimed, the limits of language provide the limits of the world. Wittgenstein referred to the whole of experience, including feelings such as pain, but from Edelman it can be seen that a more accurate view would be one which discriminated between collective, social experience and internal episodes of an evolutionarily more primitive origin.

Once we begin to do this then it becomes necessary to distinguish between different aspects of mental activity. From Edelman it can be readily appreciated that certain actions are selected for because, having been acquired, they impart survival advantages to the individual. Indeed, certain behaviours, such as drinking and feeding are essential to survival. In the absence of these behaviours the single individual could not survive for long. Other behaviours, such as those which give rise to procreation, are also essential, but this time on a species level. Without these humankind cannot survive. It may not be fashionable to do so, but it seems reasonable to term these forms of behaviour *instincts* in that they are evolved acts which serve a primary function with respect to survival. This does not mean that such behaviours are identical for each individual, but rather that their presence in some form is a species level requirement and that such behaviours exist prior to conscious awareness. As Plotkin has argued, instincts are:

> an adaptive behaviour or pattern of behaviours that is caused partly by genes and partly by the complex sequence of developmental events and experiences normally encountered by the individual members of a species. The result of such a complex genetic-environmental interaction is a brain that is structured to give rise to certain species-typical behaviours – the instincts.
>
> (Plotkin 1994: 130)

Of course the reader may recognise in this a reflection of the work of Freud, for whom instincts were central to the nature of humankind. For Freud the instincts, the drives, provided the motivational force or energy of mental activity. Though he held that the instincts were many in number, in his early work he broadly categorised these into two forms; the erotic, sexual instincts and those which related to self-preservation, such as hunger and thirst. Later he merged the erotic and self-preservation instincts into an instinct for life (Eros) and others such as aggression into an instinct for death (Thanatos). Of course Freud's theory of instincts has been questioned, particularly because, as with a great deal of Freudian theory, it has been argued that the existence of primary entities such as instincts cannot be proven empirically. I would argue that this is only the case if the concept of natural selection is ignored. As has been argued above with respect to the evolution of the human mind, once consideration is given to species level requirements for continued reproduction, then it follows that certain behaviours must always be present in a large enough proportion of the species for its survival to be maintained. If thirst or hunger caused no discomfort for the majority, and if a scx drive were not frequently irresistible for a significant proportion of the population, then humankind would not have evolved successfully.

Look at it as a design problem. You have been given the task of designing a new biological species which, once developed, you must release into the nearby woodland. To make your task easier the species will be self-replicating, so you do not have to worry about patterning a sex life for your creation. However, you do need to consider nutrition and the avoidance of harmful external influences. So after some careful thought you decide as a basic minimum the creature must be able to detect and wish to consume water, to detect and wish to consume organic material and to avoid extremes of heat and cold. At your first attempt you decide that one of the best ways of solving your problem would be to define a range of temperatures within which the creature will be constrained. Within these constraints the creature can go anywhere and do anything within its repertoire of behaviours. Next, you implant a gene that makes the creature consume water every time it is encountered. You do the same for vegetable matter. Finally you give the creature a randomised pattern of movement. Having completed this work you let 100 of these creatures free in the local woodland. After a short time (the creature has a very fast metabolism) you go to the wood and find 50 per cent of your creatures have died of thirst and 50 per cent of hunger. You realise you need a feedback mechanism of some type such that once the creature has consumed a given percentage of its body weight of water or food it stops consuming and moves off again at random. You release a further 100 and go back again after a short time to see what happened. Once again a large number have died of either thirst or hunger as by chance some never found the food and/or water (the exact number would depend on the number of food and water sources and their position and distance in relation to the release point).

You realise that one way of solving the problem would be to increase the number of food and water sources, but you know that this would also increase the number of competitors for the resources within the release area. So instead you decide that in addition to a feedback system which 'informs' the creature when it has consumed enough food or water, you also need to introduce a mechanism whereby the creature can 'smell' food and water. So the creature now has a physiological requirement for water and food, a randomised movement pattern and the ability to 'smell' both food and water. Feeling rather pleased with yourself you release some more into the woods, only to discover with this third experiment the same result as the second; some still die from thirst or starvation. Obviously, you realise at last, you need to *make* the creature go to both water and food, so you introduce a directive mechanism which will force the creature to search for both food and water. Rather than being the result of random encounters you fix it so that encounters are *driven* within the creature. Once you have done this the survival rate greatly increases and, apart from losses due to predators, your creation lives happily ever after fulfilling its given destiny as a leaf-eating, water-consuming life form.

In this simple tale of creation you have played the part of evolution and the woodland has been the selective mechanism. More usually the woodland would act as a selective mechanism on naturally occurring phenotypic variation between individual creatures, as discussed in chapter three. But what is important to realise from this simple example is that once theorising about instincts is carried out within the context of Darwin's theory of natural selection it then becomes evident that instinctive, pre-conscious behaviour associated with actions which enhance continued physical survival at the individual or species level must have evolved. If this is so, and I cannot see how this can be easily refuted, then Eros is alive and well as a contribution to understanding human nature, though perhaps the fate of Thanatos is less certain!

The example given above serves some purpose in drawing out the relationship between survival, instincts or drives and natural selection. But it is not so clear how we can use this to further our understanding of human nature. Even those who agree that instincts do exist in the manner and form described here will still be left feeling that this says nothing which relates particularly well to humankind. Humankind does have instincts, but these are ontologically no different from the instincts of other animals. This is unquestionably the case. Instincts as depicted here are a general phenomenon within the animal kingdom. Termite behaviour is instinctive as is certain human behaviour. But some animals do not only act in an instinctive way, they can also reason. And this too arises within the context of natural selection.

For a moment let me return to our creature, who during our last visit was happily consuming leaves and drinking water until it was eaten by a predator. Now, although our creation is doing quite well and its numbers are stable around a mean number of approximately eighty, we are not content any longer and we decide that we would like our creature to become more successful and double or treble its numbers. The problem is that it is a fairly easy target for any passing hungry predator. Having decided to help out our dependent creature we resolve to implant genes which enable it to recognise when any moving thing approaches, and to take evasive action. However, in consequence, the creatures spend so much time running away from each other they begin to die off once again from hunger and thirst. Over time we find there are so many options which need to be covered to ensure that all real predators trigger an evasive response, but no energy is wasted avoiding remote or absent threats, that the task becomes impossible. We eventually conclude that what we need to do is enable the creature, as a species, to learn what is dangerous and what is not dangerous. That is, to adapt to dangers as they arise. So we, as the creature's personal evolution engine, implant a desire to avoid physical damage (pain would do) and create a capability to map those occasions when physical damage has occurred. A conjunction between a 'pain map' and an ongoing event in the external world then triggers an avoidance response. In many ways the creature has the ability to remember in continuing time species-related external negative events or occurrences. As you will recognise, this is in simplified terms what Edelman has referred to as primary consciousness; the 'remembered present'. So your creature now has the ability to categorise perceptual inputs, remember them, relate these memories to criteria aligned to survival (indeed, values) and act upon the consequences. So it can now perceive, remember and learn; act upon past events to safeguard the present. And also to do this in relation to species-level values. Its actions take place intentionally with regard to evolved criteria. If this capability can evolve through natural selection with respect to predators, then it is not difficult to accept that this process may generalise to other behaviours, such as favouring some more nutritious foodstuffs over others. Although the creature started out with an instinct to satisfy hunger, through further processes of evolution it seems to be the case that an organism can eventually become discriminating.

Such discriminating acts arise from intentional mental states founded on memory, learning and species-related value. The mental map from which acts arise at a given moment in time comprise a very large number of feelings, experiences and sensations. As noted above (p. 59) Edelman (1992) refers to these directly as phenomenal states which he terms 'qualia'. Now, as we know from Edelman's theory of the morphogenesis of the brain, qualia will be different for each individual. That is, the inner mental world which is shared at the level of species nevertheless varies within these constraints at the individual level; by virtue of the uncertainty of specific outcomes inherent in the morphogenesis of the brain, as depicted by Edelman. None of us sees the world in exactly the same way; phenomenal states which inform the way we intentionally act within the world and which are directed towards the present moment (which Edelman defines as primary consciousness) are highly individual.

Now, if intentional acts directed towards the fulfilment of value criteria involve perceptual categorisation, memory and learning then, as has been argued above, such acts are not instinctive in the sense that they are an unchanging fundamental mental category of response. Nor can they be considered to be the outcome of randomised mental processes; they are instead intentional. Such acts arise from the resolution of complex mental processes which are directed towards the enhancement of individual well-being (recall that natural selection acts at the individual level and not on species as a whole). Now, although there may be no self-conscious awareness of the processes of this mental resolution, they are thus nevertheless reasoned acts – acts decided upon because they lead to preferred outcomes. But why choose the term 'reasoned' as opposed to, say, 'rational' as Plotkin (1994) has done? The distinction between reason and rationality is made to describe such acts in order to discriminate between value-directed intentional acts which arise from phenomenal states comprising primary consciousness (reason), and self-conscious action which arises from higher-order consciousness (rationality).

RATIONALITY AND THE SYMBOLIC ORDER

So what is rationality? Although there are many different authors who have produced varying interpretations and definitions of rationality, every attempt to describe what is meant by this term tends to share one thing in common: that rationality is collective property. To call a statement or proposition rational is to make a claim for its collective acceptance. Rationality is not individual in the sense that a statement can be held to be rational even if it is only understood by one person; there must be collective agreement. Thus, rational statements are statements about the world which are grounded in a collective logical or scientific discourse. A rational discourse can be termed such because it makes verifiable statements about specific properties of the external world; that is it comprises either logical or empirical accounts. As Habermas has argued:

> When we use the expression 'rational' we suppose that there is a close relation between rationality and knowledge. Our knowledge has a propositional structure; beliefs can be represented in the form of statements. I shall presuppose this concept of knowledge without further clarification, for rationality has less to do with the possession of knowledge than with how speaking and acting subjects *acquire and use knowledge.*
>
> (Habermas 1984: 8; italics in the original)

This is an important definition in that it grounds rationality within a language community. As Habermas said, rationality has less to do with having knowledge than with the way it is acquired and used by *speaking* subjects. This is not a unique standpoint; the reader is invited to compare and contrast this view with that of other key writers such as Weber, for whom rationality is also a logical and scientific collective discourse.

From the very beginning, then, this rules out the kind of knowing acquired through the processes of primary consciousness. This latter form of knowing is, as has been argued above, anchored in the past and acted upon only in the present as a kind of remembering which has no self-conscious awareness or plan for the future. In fact Habermas goes further and declares that:

> We can call men and women, children and adults, ministers and bus conductors 'rational' but not animals or lilac bushes, mountains, streets, or chairs. We can call apologies, delays, surgical interventions, declarations of war, repairs, construction plans

or conference decisions 'irrational', but not a storm, an accident, a lottery win, or an illness.

(Habermas 1984: 8)

Thus rationality is a property of humankind and not animals. If this is so then there must be something within humankind that is not present in other animals so that rationality can exist for us and not, at least in the present, for other species. And what that may be is the evolution in humankind of a higher-order, symbolic consciousness which is lacking in other animals through the absence of distinct brain structures, as has been outlined above.

As has been argued from Edelman, higher-order consciousness came into being through the acquiring of the ability to interact with the external world symbolically; that is through the generation of abstract (phonological based) symbols about the present which could be remembered and utilised to construct ideas about the future. And as humankind is biologically a collective species then this phonological symbolisation acquired a collective identity. Phonological symbols thus became shared and formed the basis for the emergence of a language. In fact it is possible to assume that if humankind had not been a collective species then language, or indeed higher-order consciousness, would never have evolved as its presence in a solitary species could impart little survival advantage. Whereas, for a collective species, a shared language can give rise to greater cooperation, thus engendering more efficient and effective hunting of prey, gathering of roots or berries or avoidance of predators.

So, rationality is the property of speaking human subjects, and humankind is a collective species which has a shared language which reflects a particular symbolic form of mentation. Rationality is thus only to be found in the symbolic order manifested in the language community. The argument is, then, that instinct, reason and rationality are quite different categories of knowing, of remembering and learning. Instinct is the most primitive in evolutionary terms in that it provides the basis of a first-order response to the external world: eat, drink, replicate. Later there evolved reason, the inner knowing which is uniquely grounded in a primary consciousness that is infinitely variable at the level of the individual. This knowing is based on value, memory and learning but it is limited to a recognition of the present. Primary consciousness makes possible an infinite range of responses to dangers and rewards not yet encountered, but being non-symbolic it does not provide a basis for acting on an anticipated future. Third, there arose higher-order consciousness through the acquired ability to symbolise the external world. This engendered the capability to formulate a concept of the future from which awareness of a temporally defined self arose. This symbolic ability is grounded in a vocalised, spoken language which came into being collectively within the human species. Through the capability to foresee came an ability to plan, to order the external world, to make rational plans directed towards achieving desired aims. This, of course, imparted great potential for humankind to construct the world in a way more favourable to the evolved needs of the species, but it would not stop there and it would be the case that, once acquired, the ability to talk about the external world would expand to include all that could be perceived, and even that which could not. As argued in Dean (1990), it is difficult to imagine that language would be restricted only to those aspects of experience which were of a practical nature. There must have been a point in the evolution of symbolic rationality when speculation took place with regard to all that could be perceived, and, where logical inferences could not be made, the remaining gaps in knowledge were invested with speculation and supposition. Indeed, in the course of time the external world became collective, social property by means of the universal propagation of language. In addition to the subjective, inner reason formed by the value-centred processes of primary consciousness, the external world became known through rational discourses. As Edelman notes:

> Once a self is developed through social and linguistic interactions on a base of primary consciousness, a world developed that *requires* naming and intending. This world reflects inner events that are recalled, and imagined events, as well as outside events that are perceptually experienced. Tragedy becomes possible – the loss of the self by death or mental disorder, the remembrance of unassuageable pain. By the same token, a high drama of creation and endless imagination emerges.
>
> (Edelman 1992: 136)

This raises an important issue: the realisation of rationality at the collective social level through acquiring language is not the starting point for the ultimate realisation of an ordered utopia, but instead may herald the loss of individual well-being. The evolution of higher-order consciousness and the emergence of collective rationality did not entirely supplant primary consciousness and inner reason in informing behaviour. Although science fiction writers often anticipate a time in the future when humankind has evolved beyond the constraints of existing biological structures, this has not yet taken place. In consequence, humankind remains subject to instincts and inner, subjective and value-based reasoning which frequently diverts human enterprise from proceeding according to the dictates of science or logic. Indeed, according to Jacques Lacan, entering into the symbolic order through the acquisition of language is one fundamental cause of the human malaise. As Lemaire has maintained, Lacan's perspective holds that:

> The philosophy which may be derived from the study of language will lead Lacan to promote the thesis that birth into language and the utilization of the symbol produce a disjunction between the lived experience and the sign which replaces it. This disjunction will become greater over the years, language being above all the organ of communication and of reflection upon a lived experience which it is often not able to go beyond. Always seeking to 'rationalize', to 'repress' the lived experience, reflection will eventually become profoundly divergent from that lived experience. In this sense, we can say with Lacan that the appearance of language is simultaneous with primal repression which constitutes the unconscious.
>
> (Lemaire 1977: 53)

What Lemaire means here is that Lacan has proposed that accessing the symbolic order places constraints on the ability of the individual to live out in the present behaviours directed by the unconscious – most usually depicted as those of a sexual nature. For Lacan, acquiring language forever objectifies primal sexuality, and from being a sexual organism we become a consciously aware being for whom sexuality is forever constrained. As Lemaire has so poetically asserted:

> All that remains of desire, of natural reproduction, of the physiology of bodies is symbols, laws, concepts or even ideologies. Marriage, family, stereotypes of heterosexual relations, fidelity, etc., are, at the symbolic level, the inevitable reductive and partially arbitrary preferential crystallization of lived biological and physiological experiences which are infinitely numer-

ous and which are henceforth inaccessible as such. For, on the one hand, the child has his vital experience in the melting pot of a culture and in accordance with its norms and, on the other hand, the symbolic catches him unawares in its nets, short-circuiting any possibility of a naïve return to the roots of his soaring flight.

(Lemaire 1977: 58)

What this means is that the content of instinct and reason as defined here may not be represented directly, consciously, in the symbolic order. Our rational, conscious self can only approximate instinct and the outcomes of primary consciousness. And this approximation can be the cause of feelings of unease, distress, frustration and even illness. This approximation arises because within the symbolic order there is a drive to achieve collective understanding, that is, the rationalisation of human affairs – the logical and scientific ordering of the world. Within a collective discourse there is only a qualified capacity for individual reason. This is so because the language community within which each of us achieves conscious awareness exists *prior* to the individual, thus placing external constraints on what aspects of our individual, subconscious self can be consciously accessed and acted upon in a knowing way In this sense, the moment we acquire self-consciousness is also the moment our alienation begins. We become in part divorced from our self.

In the movement whereby the child in one form or another translates his need he alienates it in the signifier and betrays its primary truth. The real object of lack, of need and of the instinct is lost forever, cast into the unconscious. The subject is divided into two parts: *his unconscious truth and the conscious language which partially reflects that truth.* This is also the reason for man's radical inability to find anything to satisfy him.

(Lemaire 1977: 163; italics in the original)

It is thus being argued that at some unknowable time in the past humankind evolved to a stage whereby mental life became discontinuous. Although instincts and inner reasoning continued to be represented within higher-order conscious activity, lived experience became objectified. Many aspects of human existence came to be lived not as they were in and of themselves, but through objective representations of some shared characteristics. I could struggle on trying to say this in many different ways in the hope that the central consequence of these events would become clear and still fail, so instead I will quote from C. S. Lewis who, in *That Hideous Strength* has already described the experience of a pre-conscious self beautifully:

> Mr Bultitude's mind was as furry and as unhuman in shape as his body. He did not remember, as a man in his situation would have remembered, the provincial zoo from which he had escaped during a fire, not his first snarling and terrified arrival at the Manor, not the slow stages whereby he learned to love and trust its inhabitants. He did not know that they were people, nor that he was a bear. Indeed, he did not know that he existed at all: everything that is represented by the words *I and Me* and *Thou* was absent from his mind. When Mrs Maggs gave him a tin of golden syrup, as she did every Sunday morning, he did not recognise either a giver or a recipient. Goodness occurred and he tasted it. And that was all. Hence his loves might, if you wished, be all described as cupboard loves: food warmth, hands that caressed, voices that reassured, were their objects. But if by a cupboard love you meant something cold or calculating you would be quite misunderstanding the real quality of the bear's sensations. He was no more a human egoist than he was like a human altruist. There was no prose in his life. The appetencies which a human mind might disdain as cupboard loves were for him quivering and ecstatic aspirations which absorbed his whole being, infinite yearnings, stabbed with the threat of tragedy and shot through with the colours of Paradise. One of our race, if plunged back for a moment in the warm, trembling, iridescent pool of that pre-Adamite consciousness, would have emerged believing that he had grasped the absolute: for the states below reason and the states above it have, by their common contrast to the life we know, a certain superficial resemblance. Sometimes there returns to us from infancy the memory of a nameless delight or terror, unattached to any delightful or dreadful thing, a potent adjective floating in a nounless void, a pure quality. At such moments we have experience of the shallows of that pool. But fathoms deeper than any memory can take us, right down in the central warmth and dimness, the bear lived all its life.
>
> (Lewis 1990: 670)

It is hard to imagine a better description of how an inner life independent of rationality might be experienced. What would exist would in effect be a flow of value-dependent consciousness unburdened by logical boundaries framed within the external world. And, from an evolutionary perspective, ours was the bear's experience at some time in the distant past. Before the evolution of higher-order consciousness humankind's lived experience would have been as immediate and non-reflexive as that described for the bear. The bear, behaving according to a value-dependent primary consciousness, would thus have lived a reasoned, intentional life; intentional and reasoned in that it would be directed towards the achievement of evolved, survival-related criteria. But once higher-order consciousness evolved through a series of events which gave rise to symbolic memory and ultimately a shared language, intentions and reasoning directed towards such evolved criteria became subject to rational appraisal and hence subsumed to an unknowable degree within a collective discourse. This is what Lemaire means when with respect to desire she states that:

> All the objects of the subject's desire will always be a reminder of some primal experience of pleasure, of a scene which was lived passively and will always refer back through associative links, which become more complex and more subtle with the passage of time, to that lived experience.
>
> (Lemaire 1977: 164)

The golden syrup becomes a treat, an indulgence, a gift, it will make us fat, rot our teeth, lead to a heart attack, it is a secret desire we must tell no-one about, something we enjoy which we must then vomit away, an objective category of behaviour we have more or less control over, but it is never just Goodness happening. The emergence of rationality eclipsed individual reasoning and left a void. Of course the syrup you eat still gives pleasure because the act is a reflection of an evolved value (desire), but it is a qualified pleasure:

> the subject, articulated with language, alienates his primary unconscious desire in the signifier [in effect, the collective symbolic order]. But this alienated desire does nevertheless reflect the truth of his unconscious desire and does in some way satisfy it with a substitute (the fetish for example).
>
> (Lemaire 1977: 170)

It is important to realise that desire, evolved value, is alienated rather than negated in rationality. But, nevertheless, what we experience of instinct and reason within the daily passing of the rational order is only a partial and reduced reflection of a 'pre-Adamite' human being.

So far, then, there has been developed a detailed discussion of human nature which argues, in agreement with the psychoanalytic writers Freud and Lacan, that human consciousness is characterised both by subconscious categories of mental activity which exist prior to the emergence of self-consciousness, and self-consciousness itself. Human beings have a self-aware conscious life lived out in the external symbolic order of language and rational discourse, and a subconscious life of instinct and reason structured through certain, species-specific, evolved requirements for survival. Instinct and reason can be realised, actualised, through conscious activity, but only in part. Instinct and reason, characterised for simplicity as desire, become qualified in rational life through the processes of ordering, classifying, defining, interpreting and articulating which constitute rational participation in the collective social order. In essence, this means that humankind is eternally alienated from its continuing primal nature. If this is so, then the social life which is based upon this depiction of a collective, rational self which is alienated from instinct and reason must feature essential dysfunctions. It must be the case that human social life as it exists after the emergence of higher-order consciousness is as much the source of individual malaise as it is collective success. This would suggest that we can perhaps only partly celebrate our composite self within ordered society. The promise of a utopia based on science and education promised by the Enlightenment may have proved a false hope!

Chapter 5

Reason, rationality and individual action

The preceding discussion of neurophysiology, genetics, experience, the morphogenesis of the brain and the evolution of the mind should have illustrated beyond reasonable doubt that each individual is unique in terms of their conscious awareness. Although it has been proposed that we share a primal instinct, certainly Eros and perhaps Thanatos as well (the need to kill prey and predators comes to mind), and higher-order consciousness (being grounded in the collective symbolic order) it is proposed that this is not so for primary conscious. Primary consciousness stands as the differentiating middle ground between these two other aspects of mind which both pre- and post-date its emergence. Primary consciousness is based upon value-memory anchored within a remembered present. It involves the integration of perceptual categorisation, memory and learning from which arise actions intended to facilitate survival of the individual and the species (involving, amongst others, the satisfying of hunger and sex drives). In this respect, primary consciousness functions intentionally, directed by a learned capacity to satisfy instincts in an efficacious manner. Primary consciousness is where individual consciousness begins. If it was not for the fact that the formation of primary consciousness is malleable at each moment of its genesis, then there could be no self-conscious self which harboured individual meaning. As the formation of primary consciousness progresses, through selective cell death, the selection of neural pathways by external and internal events, reentrant mapping and the evolution of global maps fashioned by continuous experience of the external world, a distinct way of seeing and valuing the world arises. As Edelman has argued, qualia, such as colours or tastes, will be not be the same for any two individuals. Differences in perception arise from differences in neural pathways and neural maps. Some differences will arise by chance during morphogenesis, whilst others will emanate from differences in experience. As discussed in chapter two, Plomin and others have demonstrated that even identical twins will be subject to differences in experience, even within the home. Instincts, which in biological terms may be considered to be the broad parameters of value resident within the limbic and brain stem regions of the brain, are given by species level constraints, as even small deviations may result in an inability to thrive. But primary consciousness is an adaptive and malleable element of the mind from which differences in meaning and reason arise. It cannot be otherwise, or, in the absence of the diversity within the species which exists at the level of the individual, natural selection could not give rise to evolution in ways shown to be the case through the work of Darwin, and many others since.

Now, so far, I have been using the word 'reason' when referring to primary consciousness and 'rationality' when depicting scientific or logical discourses which arise collectively within society. The purpose behind this distinction has been to differentiate between mental processes which arise from positions of individual value and meaning, and those which, though influenced in part by the latter, are the property of collective symbolic speculation and contemplation. The need for this distinction is twofold: it is necessary, first, to comment on why the Enlightenment promise of a social utopia has not been fulfilled, and, second, to explain why a rational order of affairs subjected to external influence is not always to be preferred. Indeed, some sports psychologists have referred to the attraction of uncontrolled and transcendent moments of experience as the pursuit of vertigo. Others depict the attractiveness of risk in over-organised lives. Similarly, with respect to the experience of a non-rational self, tennis players ascribe the term 'zoning' to those moments in a match when conscious self-awareness is lost as the game and the player merge into a seamless world of action. The so-called runner's 'high' is yet another example of this experience of the seeming dissolution of the self. Some less active people achieve related experiences through intoxication: although having experienced both the effects of too much alcohol and consciousness-shifting episodes whilst mountaineering, skiing and fell running, the 'natural high' achieved through physical pursuits is, to me, preferable. But, of course, this is a personal thing (if it was not so, it would be not be necessary to write this book).

Referring back to C. S. Lewis's creation for one moment, the pull of the bear may be within us all.

THE REASONING BEING

The separating out of reason and rationality is not new so far as sociology is concerned. C. Wright Mills commented on the division of reason and rationality in the modern world in his influential text *The Sociological Imagination*, first published in 1959. In this work Mills drew attention to what he felt was the loss of the promise of the Enlightenment as rules and procedures of the rational industrial world became increasingly beyond the reasoning abilities of ordinary people and, in consequence, individual freedom became reduced rather than expanded:

> our major orientations – liberalism and socialism – have virtually collapsed as adequate explanations of the world and of ourselves. These two ideologies came out of The Enlightenment, and they have had in common many assumptions and values. In both, increased rationality is held to be the prime condition of increased freedom. The liberating notion of progress by reason, the faith in science as an unmixed good, the demand for popular education and the faith in its political meaning for democracy – all these ideals of The Enlightenment have rested upon the happy assumption of the inherent relation of reason and freedom.
>
> (Mills 1970: 184)

What Mills is arguing here is that the individual ability to reason is not enhanced by the spread of rationality. Indeed, despite the promise of scientific discourse to render nature more subject to the manipulation of humankind (and thus impart greater freedom), the end result is that collective rational action becomes separated from the individual. Individual reasoning finds no point from which the unique interests and values of individuals, which themselves come into being by way of primary consciousness, can adequately be represented. Implicit in Mills's ideas is the notion that an essential characteristic of freedom is the means to achieve meaningful, reasoned engagement with the external world. He argues that this is lost when rational discourse goes beyond the ability of people to reason about their inherent nature. As rational action in the modern world, being grounded in science and logic, is distanced from individual reason and meaning, it is, then, alienating. The reader will no doubt recognise a parallel here with Lacan's ideas of the alienating nature of the symbolic order; experienced through the objectification of needs into an external signifier which then takes the place of the real object of desire. Although the language is very different the message is the same. In accessing the symbolic order and subjecting the external world to collective discourses which provide a logical ordering of the world according to formal laws of relationships, humankind has provided the means of its own denial of individual, specific and unique reasoning. As Mills has stated:

> The underlying trends are well known. Great and rational organizations – in brief, bureaucracies – have indeed increased, but the substantive reason of the individual at large has not. Caught in the limited milieux of their everyday lives, ordinary men often cannot reason about the great structures – rational and irrational – of which their milieux are subordinate parts. Accordingly, they often carry out series of apparently rational actions without any ideas of the ends they serve, and there is the increasing suspicion that those at the top as well – like Tolstoy's generals – only pretend they know.
>
> (Mills 1970: 186)

Of course the reader will be aware that Mills was writing about what he perceived as a crisis in society: the alienation of humankind from its own constructions. He was not overtly developing a theory of human nature. Nevertheless reference to his work is important because he separated out reason and freedom from rationality in a way that draws attention to the different nature of individual reasoning and collective rationality. Because something is rational it does not mean that it is directly related to individual human values as they have evolved and are represented within primary consciousness. Rationality, as discussed in chapter four, came into being with the emergence of a higher-order symbolic consciousness which evolved after primary consciousness arose. If a rational order (or institution) is unrelated to certain human values, if such an entity stands outside this aspect of human existence, then it cannot be a means for fulfilling such inherent values – it is not a means of expression of our inner self. Even Mills's use of the term *freedom* is consistent with this view as, whatever freedom is, it must incorporate an idea relating to the means to fulfil that which we are: the expression and realisation of our evolved being. Rationality and reason stand apart and the ability to fulfil ourselves is delimited. Indeed, if rationality and reason were inseparable then our minds would be as ordered as a formal railway timetable. This is not the case, and railway timetables are unintelligible to many people.

I have been fairly emphatic so far in insisting that reason and rationality are separate aspects of human thought. This has been necessary so as to enable the distinction between the two to be made as clear as possible. But of course in practice this is only partly true. As is described by Edelman, higher-order consciousness did not emerge as a category of mind independent of preexisting structures. At each stage of the evolution of a higher-order conscious ability, new structures were closely integrated with those already present in the brain.

In this respect both instinct and reason can be expected to be expressed in some form in higher-order mentation. As Edelman has stressed, it is the interaction between phonemic and symbolic memories and value-category memory which allows for the modelling of the world; that is, the cognitive transcendence of the present which leads to conceptions about the past and the future. Clearly, then, reason plays a part in rationality, at least at the level of value. What this means is that representations of value, of individual meaning, will be found within the symbolic order. Within the collective symbolic order there will be embedded value-positions which stem from the evolved value-criteria expressed within primary consciousness, that is, within instinct and reason. For example, sex activity conducted by means of behaviours which broaden the gene pool encourages diversity and the survival of the species. A sex drive is an instinct which is given particular expression by way of value-memory and primary consciousness such that some behaviours are chosen, valued, over others by virtue of the advantage gained by the species from this having being selected for in individual human beings. In terms of the requirement for increased diversity at the individual level depicted in Darwin's theory of natural selection, incest is less good than procreation undertaken with a distantly related person. This being so, then this universal human interest in limiting incest would, due to the nature of the genesis of higher-order consciousness outlined above, be expressed in some form within the symbolic order. This view is not as unusual as it may first seem. Many anthropological authors such as Lévi-Strauss have sought to uncover the universal character of social practices (for example, the incest taboo), which, it has been argued, occur in some form or another throughout human societies. And Durkheim offered a perspective on ethics which argued that all forms of morality were to be found collectively within society. If social life were to disappear, he argued, then all moral life would disappear as well (see Durkheim 1933). For Durkheim a person's morality, their ethical being in the world, was a product of the social history, institutions, traditions and conventions which regulate behaviour. Both these writers, in different ways, sought to uncover the way deep-seated values were both acquired, expressed and moderated within society. For both, values, whether about ethics or sex behaviour, were to be located in part within a collective discourse. Of course Durkheim was in part a social determinist in that in his relativist position (in his work he entertained the possibility that morals are epoch bound) he underplayed the importance of human biological nature in informing collective action (an error avoided by Lévi-Strauss), but nevertheless the importance of his work in drawing attention to the way in which evolved value positions may be socially embedded should not be overlooked.

Once again the reader may be reminded of Lacan's position, for whom primal desires are simultaneously suppressed and (mis)-represented in the symbolic order. Taking these views together, the case for saying that evolved values find expression in a collective discourse finds some support. But it is necessary to go further, because the implications of this eventuality are of some consequence. If primal values are represented through value-category memory in collective discourses, then rational thought, the logical and scientific pursuit of knowledge about the formal arrangement of the external world, must also be subject to limitations that arose during evolution. That is, if our consciousness has evolved from natural selection, then our consciousness at every level will mirror the concerns of survival; or at least be subject to certain limitations in this regard. How we arrive at conclusions about the world and how we solve problems will be moulded by natural selection.

The nature of human reasoning has been subject to extensive research for many years and it has been recognised increasingly that our ability to reason is closely affected by the content of the problems we are faced with. Within this emerging area of research the role of evolution in forming human reasoning has been widely discussed, and some researchers have put forward a theory of Darwinian algorithms to explain why human beings solve some types of problems better than others. However, before this research is discussed in depth it is first necessary to introduce the work of Peter Wason.

NATURAL SELECTION AND HUMAN REASONING

Wason (1972) sought to examine the way people solve abstract logical problems. In testing deductive reasoning, he devised a procedure whereby subjects were presented with an arrangement of cards with a letter on one side and a number on the other. A card could show either a number or a letter. The subjects were then asked to choose which cards they would need to turn over to determine whether the statement 'if a card has a vowel on one side then it has an even number on the other side' was true or false (that is, they had to determine if the experimenter was lying in making the statement). In formal terms, they were being asked to determine if the conditional rule *If P then Q* was being violated. The correct response in the above case would be to choose only cards which displayed a vowel (P) and cards which did not show an even number (not-Q). Other responses would be to select a consonants (not-P) and even numbers (Q). However, the subjects' responses did not follow this pattern, and Wason reports that although no subjects selected cards displaying consonants, only a minority selected cards displaying odd numbers. The most common response was, thus, to select cards with a vowel and cards with an even number (P and Q). I suspect the reader will need to think carefully to prove to themselves that the correct response is P and not-Q. What is important to discover is whether or not a vowel ever appears without an even number. There are only two possible cases that this rule was not being followed: if a *vowel* ever appeared *without* an *even* number, and if an *odd* number *ever* appears with a *vowel*. So, you would need to check cards with vowels (P) and cards with odd numbers (not-Q). The other two cards are irrelevant so far as testing the rule is concerned as it does not matter how consonants are matched; *the rule says nothing about consonants.* Consequently, selecting a consonant (not-P) or an even number (Q) proves nothing.

From this research it would seem that human beings, at least in general, have difficulty with deductive reasoning involving P and not-Q tests of validity. However, later research suggests that this is only the case with respect to abstract problems. When such tests are constructed in ways which are *meaningful* to human subjects then they are solved successfully in the majority of cases. Plotkin (1994) cites a study carried out some ten years ago in the USA. In this study subjects, who were asked to imagine they were night club door security guards, were presented with cards which depicted situations they might encounter in their fictional role. They were asked to determine which cards represented situations they would have to investigate further to ensure that the law, which in the state of Massachusetts prohibits drinking in their club by persons under the age of 20, had been broken. The cards depicted drinking beer (P), drinking coke (not-P), more than 20 years of age (Q) and less than 20 years of age (not-Q). Seventy-five per cent of subjects responded with the correct answer, that is, with P (are all beer drinkers over 20) and not-Q (are those under 20 years of age drinking beer). As the reader will no doubt readily appreciate, neither the age of coke drinkers (not-P) nor the drinking habits of those over 20 years of age (Q) matters. What these findings appear to illustrate is that human beings are better at solving meaningful problems than those of an abstract nature, at least within Western society.

There has been some considerable research effort over recent years to establish why this may be the case, and, as mentioned above, attention has been focused on the possible adaptive nature of human reasoning. That is, that the way we think is the result of natural selection – the problem-solving abilities of human beings are an adaptive response to the environment experienced by human beings during the evolution of rational mentation.

The issue of the extent and nature of the effect of natural selection on human mentation has been researched extensively by Leda Cosmides in a series of studies carried out in the 1980s which focused on what was termed 'the logic of social exchange'. With respect to the difficultly human beings have with abstract reasoning she noted that:

> The study of human reasoning has been dominated by the search for content-independent cognitive processes. Early research started from the premise that humans reason logically, that is, using the rules of inference of the propositional calculus. These

rules of inference are content independent: they generate only true conclusions from true premises, regardless of what the propositional content of the premise is. However, more than a decade of research has shown that people rarely reason according to these canons of formal logic. Moreover – and contrary to initial expectations – psychologists found that human reasoning is content dependent: the subject matter one is asked to reason about seems to regulate how people reason.

(Cosmides 1989: 191)

Underlying Cosmides's position was the view that human beings are not, innately, particularly good at abstract logical problem solving, and that in general our problem-solving abilities are closely related to the requirements of our primal environment. In other words, our problem-solving abilities relate to the particular circumstances of our evolution. As Cosmides maintains:

Even if they have not paid much attention to the fact, cognitive psychologists have always known that the human mind is not a computer, but a biological system 'designed' by the organizing forces of evolution. This means that the innate information-processing mechanisms that comprise the human mind were not designed to solve arbitrary tasks, but are instead *adaptations*: mechanisms designed to solve the specific biological problems posed by the physical, ecological and social environments encountered by our ancestors during the course of evolution.

(Cosmides 1989: 188; italics in the original)

Cosmides argued that Darwinian evolution led to the selection of particular forms of mentation:

Natural selection, in a particular ecological situation, constrains which kind of traits can evolve. For many domains of human activity, evolutionary biology can be used to determine what kind of psychological mechanisms would have been selected out, and what kind were likely to have become universal and species-typical. Natural selection therefore constitutes 'valid constraints on the way the world is constructed'.

(Cosmides 1989: 189)

She drew attention to the adaptive nature of human reasoning, in that she proposed that the way we think is derived from species-level requirements to solve certain types of problem. Her work focused on the rules of social exchange which, it was argued, have a particular form because of the evolved need to engage successfully in specific social interactions, one of the most important of these being given as the need to evaluate the processes of social exchange. As humankind is biologically a collective species, then, at the level of the individual, an innate ability to evaluate the extent to which the rules of collective engagement are not violated would impart a survival advantage (based on the premise that violations would cause personal disadvantage). Such an ability to detect violations would be an adaptation which arose from natural selection. To test these assumptions, that human beings have a particular facility to solve problems relating to social exchange, Cosmides undertook a series of experiments based upon the Wason selection test.

Cosmides (1989) tested the validity of the proposition that we have an evolved capacity to solve particular problems relating to specific circumstances of human existence. The characteristic of social exchange elected by Cosmides to illustrate the adaptive nature of reasoning related to rules of social exchange concerning the evaluation of cost–benefit situations, that is, the ability to evaluate that if someone in the tribe receives a benefit then they must pay a price. Clearly, in mutually dependent groups, it is important that each individual is able to ensure that others in the group do not gain a benefit from the tribe without contributing appropriately to the tribe. If this is not so then some will expend more time and energy on survival and reproduction than others; clearly a disadvantage in survival terms. In essence, such a rule would state that you do not get the benefit if you do not pay the cost. What is important, then, is to be able to spot the cheaters; those who take the benefit but do not pay the cost. That is, the rules of formal logic are less important in adaptive terms than rules which detect cheaters.

As the reader will know from the discussion above, a Wason selection test examines whether the subject can determine if a conditional rule 'If P then Q' has been transgressed. In focusing on issues relating to social exchange Cosmides (1989) selected rules of the form 'If you take the benefit then you must pay the cost', and constructed rules which, though related to social exchange, were unfamiliar in actual content. Her reasons for doing this were to test the legitimacy of 'availability' theories, which explain the results of tests such as the one quoted above concerning underage drinking as arising from familiarity. However, for the purpose of making this account as clear as possible to the reader Cosmides's original terms *have been substituted by terms which may be more familiar to the reader*. So be aware that what follows is a much abbreviated account of a complex and original research project (the interested reader is recommended to consult Cosmides 1989 directly for a full account; Plotkin 1994 also provides a useful summary of this important work).

Imagine you are the leader of a tribe which has a law which states that 'If you want a hut, then you must fell a tree' (If P then Q). How would you test if the law is being upheld? The best way to do it would be to check if those who have a hut have felled a tree (P), and if those who did not fell a tree did not have a hut (not-Q). P and not-Q, as the reader will know from above, are also the logical falsifiers of a conditional rule of the form 'If P then Q'. Using this approach, but with slightly different wording, Cosmides (1989) found that the majority of respondents made the correct (P and not-Q) response. Given that the laws used related to unfamiliar social contexts, ones with which the subjects would not have been familiar, then it is reasonable to claim that an 'availability' thesis cannot explain the results achieved. However, Cosmides took the research a stage further and created 'switched' laws of the form 'If you pay the cost, then you take the benefit'. So, in our example above, the law would become 'If you fell a tree, then you will have a hut'. The logical falsifiers of this conditional rule are the same as the non-switch law, P and not-Q. However with this form of wording the great majority of respondents chose not-P and Q. What the respondents were doing was selecting for cheaters rather than checking if the law had been broken. If, as the tribe leader, you only wished to test if the law is being upheld then all you would need to know is that if someone has felled a tree then they must have a hut; so you check P (felled a tree) and not-Q (does not have a hut). However, as Cosmides (1989) explains in detail, our capacity is, for adaptive reasons, directed towards seeking out those who get the benefit but do not pay the price: the cheaters. Hence, as tribal leader, what we would be interested in is those who have a hut but did not fell a tree. So the theory of 'looking for cheaters' would predict that not-P (did not fell a tree) and Q (did have a hut) would be selected instead of the logical falsifiers P and not-Q. In Cosmides's research this proved to be the case for the majority of subjects, thus supporting the proposition that human reasoning is constrained towards Darwinian algorithms selected for during the evolution of the human species, at least in terms of the logic of social exchange. Hence it is not surprising that we are relatively poor as a species at solving contentless abstract logical problems.

The work of Wason and Cosmides is significant in terms of both drawing out important features of human reasoning, and, in terms of Cosmides's work, in showing how in critical ways these are the result of natural selection. Thus, even when it comes to considerations about the nature of our symbolic reasoning facilities, it would seem that we are constrained by adaptive forces.

ADAPTIVE RATIONALITY AND MORALITY

In her presentation of her findings regarding social exchange theory, Cosmides (1989) speculated on the possibility of the presence of a deontic logic in human mentation; that is, the presence of inferential rules relating to obligations and entitlements. She notes that Manktelow and Over (1987) raised this question in their review paper on human reasoning, but she concluded that such a theory would not explain the findings from her research on social exchange. She acknowledged that, in a limited way, the logic of social exchange does incorporate rules relating to obligations and entitlements (looking for cheaters implies an inserted *must* regarding paying the cost for a benefit received) but argues that the 'look for cheaters' procedure only exists with respect to exchange around a cost–benefit structure. She argued that a rule which has an obligation or entitlement but lacks a cost–benefit structure will not elicit a pattern of response which signifies the operation of a logic of social exchange, as presented in her 1989 paper. I find this a reasonable conclusion as to argue otherwise is in some respect to pursue a reductionist approach within which the logic of social exchange is merely a sub-set of deontic logic, which in turn may be the sub-set of something else. Such an approach would ultimately have to uncover a final seat of human reasoning which would have to be sufficiently abstract so that all other more externally directed sub-sets of logic could be seen to stem from it. It seems more reasonable to suppose, as Cosmides argued, that human reasoning is modular. We most probably have different algorithms with different adaptive functions: distinct logic mappings which have evolved to contend with different external issues in the same way that neural maps and global mappings evolved – in response to natural selection. If this is so, that the logic of social exchange is not the only reasoning function to have evolved, then, in the way that Cosmides made assumptions about essential aspects of human life as hunter-gatherers (a state which has occupied 99 per cent of our history), other deductions can be made regarding the presence of further adaptive algorithms.

Now Edelman has made it clear that the structure of our brain is such that even higher-order consciousness is linked through neuro-physiological structures to more primitive areas of the brain associated with value. Thus the speculation has been put forward that species-level values, as expressed in the instincts and reason, would be represented in higher-order, symbolic mentation. Clearly Cosmides's work provides good evidence for this to be the case. Indeed, her work lends strong support to the proposition that reason and rationality (as depicting formal logic and science – content-independent inferential processes) are separate aspects of mentation. We may expect human beings to act rationally, but they are more often likely to respond with a rationality constrained by adaptive processes. We can solve abstract problems and, indeed, many people gain a great deal of pleasure from doing so. For example, struggling with a mathematical proof or solving the week-end crossword puzzle can be a pleasurable experience, but the evidence shows that it does not come as naturally as does judging the outcome of a social contract. But there are other judgements which also seem to have been arrived at with little difficulty even though they too involve complex situations. I refer of course to moral judgements which, following Manktelow and Over (1987) and Cosmides (1989), involve forms of obligations and entitlements other than those relating to costs and benefits as depicted in social exchange theory.

It has been argued that the logic of social exchange arose through natural selection because it imparted a survival advantage. In the same way, then, it can be argued that further Darwinian algorithms will have arisen from the same cause with respect to other aspects of hunter-gatherer life. Now one of these, it may be argued, must relate to obligations towards young children who would be unable to thrive without the continuous support of adults. It can thus be hypothesised that the logic of social exchange will operate differently with respect to children as opposed to adults. Children need to be nurtured as long as they are vulnerable, hence it cannot be expected that a 'look for cheaters' strategy would be employed in cost–benefit exchanges as to do so may place them in jeopardy. Instead it can be hypothesised that in the experimental environment of a Wason selection test, responses would show a 'look for disadvantage' strategy, as the prime interest would lie with ensuring that children receive a benefit even though they may not (may be unable to) pay the cost. Hence the expected response to the law 'If a child sits by the fire, then they must have chopped some wood' would be not-P and Q (the reverse of the 'look for cheaters' strategy) and its switched version 'If a child chops some wood, then they can sit by the fire' would be P and not-Q (again the reverse of a 'look for cheaters' strategy). The adaptive logic of the situation would be directed towards minimising the possibility of excluding *any* child from the fire.

Although this may seem unlikely, it is reasonable to suppose that a different logic of social exchange will be applied to young children than to adults. If this were not so, then young children unable to fulfil consistently the full requirements of tribal membership would be unable to thrive amidst adult competition for scarce resources, which of course is the basis of natural selection. An alternative explanation may be that judgement of the extent to which children follow the laws of social exchange is merely suspended. However, the question would still remain as to what logic informed that suspension; how did we come to decide the matter? Such an explanation as the latter would simply remove the cause of a 'look for disadvantage' strategy to a different sphere of mentation, but one which would nevertheless have to have arisen adaptively. The way we relate to children with respect to cost–benefit interactions must be adaptive in the same way that it is with respect to adults; only the outcome is different. Hence, although the example of a 'look for disadvantage' logic of social exchange given here may be unattractive, or indeed may strike you as being simply misguided, Darwinian algorithms associated with deontic logic must have arisen during the course of human evolution. This must be so as, first, it is predicted by Edelman's theory of the evolution of the brain, depicting as it does the close integration of primal and higher-order mental functions. Second, algorithms associated with the logic of mutual support and the nurturing of young must have evolved in a collective species with higher-order consciousness within which these behaviours are not only seen to exist, but are clearly and knowingly valued.

Now, the idea that value may somehow be associated with the execution of Darwinian algorithms needs some explaining. At the species level there are primal values associated with fundamental needs such as sex reproduction and hunger which, during the process of evolution, came to be expressed in individual reason as both constraints and directing elements of the delimited remembered present of primary consciousness. During the evolution of higher-order consciousness these values, and their reasoned derivatives expressed within primary consciousness, subtended the emergence of logical reasoning. What emerged in consequence of these events was a rationality constrained by the adaptive nature of human mentation. The way we think about things of a social or personal nature is informed by value-reason. Reasoning, as depicted by Cosmides, Wason and others, is, in the context of the terms used throughout this current text, a reasoned, adaptive rationality. Conversely, rationality as mentation associated with content-independent formal logic and science, is a collective discourse(s) distinct from the value and reason which evolved during humankind's previous history as hunter-gatherers. Our *reasoning* is separate from the rationality required of us during day-to-day professional encounters in the post-industrial world.

THE DISJUNCTION OF REASON AND RATIONALITY

One outcome of this is that values are simultaneously present and absent in the symbolic order. In consequence it is of use to draw a distinction between reason and rationality in a way that recognises the role of adaptation in structuring human mentation. In this sense we can view our dealings with rules which relate to obligations and entitlements in our private and personal lives as being informed by *reasoning* or, in other words, an adaptive rationality. Conversely, our public and professional selves will, in general, operate according to the dictates of a formal rationality which is not grounded in a deontic logic. The dilemma for human beings in complex societies such as our own is then one of being required either directly or from the circumstances of our daily lives to act outwith judgements derived from reasoning as an adaptive rationality (where our judgements make intuitive sense to us), and instead follow the logic of a rationality disembedded from adaptive values. In other words, we are frequently required to act in ways which are not valued or may indeed be objectionable (we are required to act amorally or immorally, in terms of our inner values). For example, in a previous paper (Dean 1990) an analysis of teenage drinking and soft drug use in the Western Isles of Scotland was undertaken to explore differences between them. During the fieldwork it had been observed that heavy under-age drinking and extreme drunkenness by young teenagers was experienced differently than involvement in illicit soft drug use. The former gave rise to few perceptions of wrongdoing by those involved, whereas for certain school-aged teenagers soft drug use gave rise to an experience of moral dilemma. In analysing these different responses it seemed to be the case that actions which were thought to be permissible derived from the participation in a normative collective celebrative event, whereas those which were not thought to be permissible, were not part of a normative social event, led to the experience of discomfort and/or forms of dissatisfaction with the self. It was further argued that involvement in activities which departed from the communal (moral) order (such as that surrounding the use of alcohol in Hebridean culture – see Dean 1995b) gave rise to personal negative consequences. It was argued that to be displaced cognitively from the shared value position of normative rituals (in this case joining with others in the normative practice of drinking alcohol) led to a certain loss of well-being, often expressed through a dissatisfaction with personal behaviour (letting oneself, one's friends or one's family down), feelings of guilt and fear of discovery and social censure. Participation in a collective celebration mediated by alcohol was, within this social grouping, a reasoned activity arising from an adaptive rationality in that the behaviour stemmed from an adaptive trait associated with collective gatherings (humankind being biologically a collective species). This was not so with the use of soft drugs, which, due to their illegal status and relative rarity in that community at the time of the research (1987–90), were consumed in secret either individually or in small groups.

I do not want to overstate the strength of the relationship between well-being and the exercise of individual reasoning or adaptive rationality, but I do want to illustrate that a certain connection does exist. This can be achieved by summarising, and thus simplifying, some possible consequences of actions which arise cognitively differentially from reason as primary consciousness, reasoning as adaptive rationality or rationality as content-independent inference. In the case of the Hebridean example, all of the actions of each individual could be considered to be *reasoned* in that they arose from the individual orientations of the people involved (recall that through the processes which underpin the morphogenesis of the mind we are all different from each other to a greater or lesser extent neurophysiologically and hence cognitively – remember that not all the subjects in Cosmides experiments answered according to the logic of social exchange), but only the normative alcohol drinkers displayed an adaptive rationality in their choice of patterns of socialising. The soft drug users' actions were not grounded in traditional forms of socialising, they were thus not grounded in an adaptive rationality which informs the nature of collective gatherings (and from the fact that their actions gave rise to a loss of well-being, it would seem that they were not grounded in other modes of adaptive rationality either as this implies valued actions).

What is important to focus on here is not the content of the actions themselves – this would vary from one culture to another – but on the presence or absence of a normative moment in each respective set of acts. The adaptive trait it is claimed is relevant here would be associated with the value of social gatherings which received collective endorsement, not with any specific behaviour. With respect to these activities alone, content-independent inference played little notable part in their actions. Indeed it could be said that, given the negative consequences of both soft drug use as described here and the health and social consequences which may stem from heavy drinking, both sets of actions were irrational, in the formal sense. Nevertheless, both were grounded in *reason* as the expression of individual meaning at the level of primary consciousness, though only drinking, it is claimed, displayed adaptive (and thus symbolic) rationality.

The reader may wish to think of other and perhaps much better examples of contemporary disjunctions between reason, adaptive rationality and formal logic. I would like to suggest that the mass destruction of human life during certain twentieth-century wars, and the continuing felling of the Amazon rain forests may be further good examples of the often unfortunate consequences which originate from the dominance of formal rationalising over *adaptive* rationality.

Chapter 6

Adaptation and uncertainty in human nature

The previous five chapters of this book have traced out a complex perspective on human nature. In arriving at the final statements of chapter five which claimed simultaneously that individual reasoning may be irrational and that rationality itself may be frequently alienating, the reader has been requested to consider the findings of research from a variety of fields, from biochemistry to the philosophy of human nature. This may at times have appeared to be an unnecessary task, or at least one which could have been simplified, with much technical detail omitted. After all, it may be claimed, it would be possible to argue for an adaptive stance on human nature without having first to consider the neurophysiological genesis of the brain. This, surely, can be left to biologists and medical scientists? Well, if that was all I set out to do, to present a theory of human nature based upon natural selection, then such claims could be held to be reasonable. But, although this is clearly one of the aims of this current book – to offer a contribution to current work in the social sciences seeking to discover and unfold the presence of Darwinian phenomena in the social world – there is an additional task towards which this book is directed. This second focus is concerned with uncertainty.

Throughout the previous detailed discussion of the physical and social genesis of human nature one important phenomenon may have revealed itself to the reader: that at each stage in the onto-genesis of the individual there occurs great uncertainty with respect to outcomes. Even with regard to the metabolism of alcohol and other mood-altering drugs individual differences do exist. A wide range of research evidence has shown that differences in the way these substances are metabolised can give rise to distinct levels of use from one person to another. Indeed, in chapter one research findings were presented which illustrated that differences in drug and alcohol consumption were both qualitative and quantitative in nature. That is, for example, the world is not divided into those whom do or do not use cocaine, but is instead comprised of people with wide-ranging differences in levels of use and time periods of use. Some use a little every day of the week and yet others will have used large amounts some time in the past. Understanding generalised metabolic pathways and sites of action of certain intoxicants thus does not allow predictions to be made with respect to specific outcomes for an individual. If use is not simply present or absent, but is instead continuously variable, as work on Quantitative Trait Loci suggests, then predicting outcomes for individuals becomes extremely difficult, if not impossible. As research on QTLs has shown, a specific quantitative trait can arise from the interaction of a number of different QTLs on the genome. Given that it is known that gene expression is affected by environmental factors, such as levels of indigenous proteins which will vary over time, then the system within which responses to intoxicants arise can be seen to be dynamic. So there exists a dynamic equilibrium whereby environmental changes which affect gene expression may subsequently change the action of a complex system of linked QTLs, which will in turn give rise to yet further changes in the state of the system. In such a dynamic system the only way to be able to predict what will happen next is to know the exact state of all contributing influences at the start of your observations, as the state of the system at time 0 will affect the state of the system at time $0 + 1$, and so on. Consequently, even if it was possible to identify each component of the brain involved in the metabolism of responses to intoxicants, we would still not be able to predict what would happen over a person's lifetime unless we could achieve at one moment in time a complete description of the state of every element. Even the lack of knowledge of one element could lead to an unexpected outcome through the existence of a level of feedback to an interrelated element we had not anticipated.

Now, although it may thus be a worthy aim to seek to achieve a complete description of the state of the brain at a given moment in time, it can be appreciated from Edelman's work that this is unachievable. As we know from chapter three, the exact outcome of the processes which underlie the morphogenesis of the brain cannot be known. As we now know, the action of natural selection on the formation of neural pathways, and subsequently neural and global maps, is such that each individual is essentially unique. Of course, as the reader will recall, there are species-level constraints on what is possible, but in important respects each brain is different and so, therefore, is each mind. Natural selection acts upon the brain at each level of its formation. During periods of cell division, movement and differentiation some cell lines survive whilst others do not. And later, during the formation of neural networks, selective processes once again lead to the propagation of some neural pathways as opposed to others. The result of this is that we all perceive things slightly differently from one another. Qualia, such as colours, tastes or smells, will be different for each of us.

Although Plomin and others working on the links between genetics and experience did not make direct reference to an uncertainty of outcomes in the development of behavioural traits, such as cognitive abilities, and environmental measures, such as academic success, their work certainly draws attention to the unpredictability of outcomes. There is, as discussed in chapter two, research evidence from a range of studies on the relationship between genetics, environment and experience which argues for differential developmental outcomes in consequence of differences in environment, even for identical twins. And other evidence, as cited above, shows that those who are raised in families where there are alcohol-related problems are as likely to abstain from alcohol as to develop alcohol problems themselves (Valliant 1983). Overall, there seems to be considerable evidence to suggest that any specific behavioural outcome, and not just substance use even though that is the substantive focus of this current book, is difficult if not impossible to predict at the level of the individual due to the complexity of the biological, selective (recall that natural selection acts upon the individual) psychological and social circumstances through which the individual comes into being.

This uncertainty is to be found at every level, from the uniqueness of the home environment through to dissimilarities in the structure of the brain. In consequence, it is acceptable to suggest that what we are as individuals is a resolution in a given moment of all the dynamic elements of being, be they biological, psychological or social, from which we are constituted. For each of us, different events and influences, which are in themselves constantly changing, contribute to our state of being at a certain moment in time. At the simplest level we can easily see this to be the case: the new neighbour who interests you in a pastime you had never before contemplated; the job opportunity you saw by chance in an evening newspaper which seemed to meet your growing need for a new challenge fostered by your existing mundane occupation; the person you now live with whom you met at the party you never meant to go to, and would not have gone to if it had not been for the bus which knocked you down and broke your leg thus causing you to cancel the skiing trip you had set your heart on. All of these fictitious events depict accurately the world we recognise: one of frequent change and unpredictability. This may seem a rather unremarkable insight, but on closer reflection these simple scenarios can reveal a more complex reality still. On some days you would read about the job and act upon it, yet on others it would seem beyond hope that you could be successful. Sometimes you would put behind you the disappointment of a lost holiday and enjoy yourself in other ways, but at other times you would perhaps stay at home and feel sorry for yourself. On the basis of visible evidence it is often difficult to determine which way a person will act. Indeed, often we are not sure ourselves how we will act in a given set of circumstances. Sometimes even the most minor comment can cause distress, whilst on other occasions we are relatively resistant to similar intended or unintended provocations.

What is being outlined here are two distinct though interrelated phenomena. The first is associated with uncertainty of outcomes, which stems from variation in development processes and differential environmental influences. The second phenomenon relates to unpredictability of responses. In other words, each individual is unique cognitively (both reason and adaptive rationality are highly individual) and changeable with respect to their orientation and responses to external events. The analysis so far has provided insights into the origin of biological and cognitive difference, but the introduction at this stage of an idea of non-permanency with respect to orientations and reactions to the environment is a new concept. However, although this may be a relatively novel idea, it is in many ways an obvious extension of the preceding discussion. Throughout most of the text so far a central place has been given to the work of Darwin and the importance of natural selection in framing human primary and symbolic consciousness. Indeed the research findings of Cosmides suggest strongly that both our ability to reason and the form such reasoning takes are the consequence of natural selection. Our reasoning powers are adaptive in that they were selected for with respect to the problems, tasks and responsibilities encountered by our hunter-gatherer ancestors. Cosmides argues convincingly that our reasoning facilities are modular in that different 'Darwinian algorithms' have evolved in response to differing external requirements. If this is so then, either potentially or in the past, there were once more than one similar (perhaps only partly formed) algorithm upon which natural selection could act. In the way that Edelman argues with regard to the selection of specific neural pathways (those which are 'successful' are promoted, and those which are less adaptive are unused), it is possible to argue that a similar process, at a higher-order level, occurred with respect to the selection of specific algorithms. There are two ways this could have taken place. First, individuals could have had more than one algorithm which was associated with a particular issue in the lives of hunter-gatherers, and in the same way that 'successful' neural pathways are selected, the one that works best was selected over those with a less good fit with respect to achieving a 'best solution' outcome. Or, second, different individuals had different algorithms and therefore selection took place at the level of the individual person.

Given that natural selection appears to display the feature of 'scaling', by which it is meant that natural selection operates simultaneously on various levels (at the cellular, neural pathway and whole organism levels), then it is likely that both differences between competing algorithms and individuals were present during the operation of natural selection on human reasoning. And, of course, there is no obvious justification for stating or implying that such processes are historical. In the same way that Edelman's natural selection of neural pathways is held to be a continuing feature of the morphogenesis of the brain, it can also be argued that the natural selection of Darwinian algorithms is an enduring feature of the formation of individual adaptive rationality. What this means is that the Darwinian algorithm associated with cost–benefit social exchange contracts has more than one form. It can be argued that this will be the active form for the majority as it can be considered both to have a form of dominance due to previous selection in the Pleistocene era *and* because it will have been subject to a form of *secondary* selection through environmental experience during childhood development (cf. Edelman's species-level constraints and the natural selection of neural pathways during childhood). The possibility of such a binary system of natural selection is to some extent hinted at in Cosmides's data; not everyone got the right answer. Only about 75 per cent responded as predicted, so 25 per cent reasoned the problem differently. This large minority deviation from the expected results suggests clearly that forms of reasoning other than those arising from the algorithm which fits best with the logic of social exchange exist to a notable extent in the population. It may be that individual subjects in Cosmides's research who calculated the adaptively less good answer were merely dysfunctional with regard to such tasks, in that their cost–benefit algorithm was somehow incomplete. However, this would be a curious situation if it were true as the argument that the cost–benefit algorithm is adaptive requires that over time algorithms which do not successfully calculate cost–benefit scenarios would be selected against and become extremely rare. An incidence of 25 per cent is not a rare event so either cost–benefit algorithms are not adaptive, which is highly unlikely, or cost–benefit algorithms are adaptive but they still exist in a variety of forms, only one of which is selected for during childhood development. During childhood the 'Cosmides algorithm' is selected for in the majority of cases as it has acquired a species-level quasi-dominant property and, importantly, it fits best with prevailing logics of social exchange.

The existence of such scaling in natural selection would enhance the adaptivity of human reasoning in that it could respond well to rapid changes in the social order. Of course there will be species-level constraints on what would be possible. We are, as has been stated a number of times, biologically a collective, social species, so, there will be limitations on the range of algorithms which may pattern human reasoning. This would particularly be the case in the era of hunter-gatherers, when mutual dependence within small groups was essential for survival and, thus, there would have been less indigenous tolerance to variation. However, in the contemporary world, where cooperation around the material reproduction of society is mediated by way of a content-independent rationality, as expressed in formal logic and science (within, say, industry and commerce), variation in individual adaptive reasoning may be less selected against in contemporary time and thus more possible. The plethora of rationalities claimed by post-modern theorists to exist in the contemporary developed world may in fact be the simultaneous expression of different forms of adaptive reasoning. There remains a collective formal rationality embodied in the institutions, traditions and precepts which, to paraphrase Durkheim, structure social action in the post-industrial world, but there also exists the adaptive rationality which, although being symbolically mediated, is grounded in instinct and reason. In consequence, the way we interpret the world and act upon it is essentially unique. We each make sense out of the world in different ways. What often appears on the surface to be uniform behaviour and motivation is unlikely to be so in practice.

The importance of focusing on individual consciousness as opposed to pursuing research which assumes that symbolic reasoning (as adaptive rationality) is a collective property is beginning to be recognised within the social sciences. This has not always been the case; too often social analyses have treated human beings as though they were a mere extension of a collective consciousness instead of being individuals reflecting their own reasoning in an intentional manner. As Cohen has noted:

> Who has the right to determine who a person is: the person in question, or those with whom the person interacts? In treating the self as socially constituted, social science has denied 'authorship' to the individual, seeing identity either as imposed by an other, or as formulated by the individual in relation to an other. Both views imply the insubstantial nature of selfhood.
>
> (Cohen 1994: 73)

The central focus of Cohen's thesis is that society, the collective rational order, does not impose a single way of seeing the world on individual members, rather it provides a setting comprising institutions, customs and traditions within which people come into being as conscious, intentional individual participants. Culture provides a symbolic structure but it does not determine meaning. This of course makes a claim for the uniqueness of consciousness, particularly with respect to symbolic reasoning. As Cohen argued:

> Symbols are cultural (therefore public) forms, the meanings of which are substantially private. Over the last twenty years, it has become a commonplace of symbolic anthropology that the meanings of symbols are not exhausted by their shared or public elements, but are essentially a matter of private interpretation and, as such, may be inaccessible to others (including, alas, to ethnographers). Metaphor is public in so far as its terms are culturally salient and compelling; but its meaning to

the different individuals who are oriented to it may be utterly different.

(Cohen 1994: 142)

This depiction of metaphor is very close in meaning to Edelman's definition of qualia. Although we may all see the colour red, or hear the word 'food', what we see and what we hear will be unique to each of us. And given that it may be that cognitive algorithms have multiple forms subjected to natural selection during early development, then even concepts, moral judgements or other symbolic representations of the material or social world will vary in meaning between individuals. Cohen provides a helpful example which draws out this point with great clarity:

> If I was a practising Jew, I would spend Yom Kippur, the Day of Atonement, in a synagogue with my fellow Jews, fasting as they all fast, reciting the same liturgy as they do at more or less the same time, beating my breast to give emphasis to the confession of my misdeeds and to ask for forgiveness and a clean sheet in the divine ledger. I would bow with them in a gesture of reverences as the scroll of law, the *sefer torah*, is carried past us in procession. With my neighbours, I would almost certainly utter a comment of relief when the *shofar* was sounded in the evening to signal the end of the fast. I would nod in agreement or acceptance during the Rabbi's sermon as he pronounced authoritative interpretations of scripture. But nothing in all of this choreographed, uniform behaviour entails that my experience of prayer, or of faith, or of the religion generally is the same as anyone else's. We may weep together or exult together, but still the meanings which religious commitment have for us may be quite different. At the very least, we cannot know that they are the same. We deploy the same symbols to signal the commonality of our beliefs, but this says little about how we interpret and make meaningful to ourselves those symbols.

(Cohen 1994: 19)

Essentially, collective participation and the sharing of events both behaviourally and linguistically do not imply an equality of shared meanings, values and interpretations placed upon those events by the self.

The whole of this book to this point has been an attempt to provide evidence in support of this position by detailing those aspects of the genesis of the self which underpin the claim of conscious individuality over collectivist interpretations of human nature. At each stage of biological morphogenesis there is randomness, diversification and continuing natural selection, all of which foster difference rather than uniformity. In consequence, the structure of each human brain is distinctive, even to the extent of there being an inherent uniqueness with respect to the formation of neural pathways. And from these differences in neural tissue arise differences in perceptions, values, instincts, reason and adaptive rationality. As has been argued, this absence of uniformity in cognition means that to a large extent we are each a distinct and separate initiator of our own being-in-the-world. But this does not imply the existence of a wide-ranging freedom of will. It is important to recall that we have achieved a particular form through the process of evolution, and in consequence there are species-level constraints upon us which structure the way we consciously engage with the external world. And it has also been argued that natural selection is involved in the processes of *post partum* development. Edelman has argued that this is so with respect to the formation of neural tissue, and in this chapter it has been contended that forms of adaptive reasoning may be selected for during childhood. That is, the induction of specific Darwinian algorithms may take place by way of natural selection at an early stage of development. Indeed it may mark the end phase of a process of selection which begins with neural pathways and ends with the selection of the neural maps which may prove to be the morphological root of cognitive algorithms.

From what we know about the morphogenesis of the brain and the role of natural selection in forming cognition it seems more than likely that our self-conscious self comes into being in the way depicted here. Each person is unique, though ultimately determined by complex processes of development. The meanings we derive as individuals from drinking alcohol or injecting heroin may be distinct to us, but they have ultimately derived from processes we are only just beginning to understand. Brain chemistry is important, and in this regard the importance of the inheritance of certain characteristics of brain function cannot be ignored. But it is equally important to acknowledge that early experience also plays a part in determining how we see the world. The fact that a given person may have a certain inherent potential with respect to intoxicants (for example, such substances may be metabolised within the brain in a way which makes the person more susceptible to continued or extensive use) does not mean that early experience, through acting directly on the formation and induction of specific neural structures within the brain, cannot give rise to other outcomes. In fact, rather than helping us to predict outcomes for individuals, the perspective being developed here implies that it is difficult if not impossible to know what will happen. Although what we may become is determined by the processes discussed here, there remains great uncertainty with respect to outcomes. To repeat the main conclusion of this stage in the argument, human existence is a complex phenomenon which is based upon the resolution in a given moment of a very large number of contributing elements.

The study of such complex phenomena has recently drawn a considerable amount of attention. Starting with studies of weather systems, where it has long been known that predictions over more than short periods are difficult to make accurately, many research scientists, including biologists, economists and historians, are beginning to articulate the foundations of a chaos theory of natural systems. Chaos theory is concerned with the characteristics of non-linear systems; systems which do not progress in an easily predicted linear (uniform) manner. Now, from what has been discussed so far it would appear that aspects of human existence could be described as such a non-linear system. Hence, in terms of advancing this current analysis of human nature and the genesis of intoxication, additional headway may be made if attention is turned towards the possible contribution of chaos theory.

Chapter 7

Principles of complexity and chaos

One thing that is obviously apparent from the preceding discussion is that reasoning, rationality and human nature are complex phenomena. It will be clear that human beings are not hardwired entities whose responses to the external world are somehow given at birth or elicited by the material and social environment. Instead, a considerable weight of research findings and philosophical reflection has revealed humankind as a product of intricate and non-uniform historical and developmental processes.

The fact that life can be unpredictable can be readily appreciated. Which of us could have predicted ten years ago that we would be doing what we are doing now in our professional and private lives. It is generally true that the unexpected frequently occurs; we know that something new will happen to us, it is just that we do not know exactly what it will be or when it will occur. But why should this be so? Why is it that the passage of our lives, of society and even civilisations appears to be unpredictable? To answer this question it is necessary to review findings from studies of non-linear dynamical systems, an area which has come to be referred to as non-linear dynamics or chaos theory.

The evidence for the existence of chaotic systems stemmed from research on weather patterns. It was a paper published by the meteorologist Edward Lorenz in 1963 on non-periodic flow in weather systems which provided the foundation for this new area of science. Indeed, citing the weather as an example of chaos is common as it is quite easy for most people, particularly those in temperate climates, to appreciate the uncertainty which surrounds weather forecasting.

The fact that it is difficult to predict the weather may be common knowledge, but the reason for this is not necessarily easy to appreciate. After all, it may be argued, weather outcomes arise from the interaction of a small number of primary elements; atmospherical pressure, humidity, wind direction and temperature. Surely, it could be argued, if we knew the values of each of these factors at a particular moment in time then predicting the weather would be a simple business. We know that in practice this is not the case and that, instead, although weather forecasting can be accurate for a few days, over longer periods of time predictions break down. For example, a recent weather forecast in a British national newspaper described the weather as continuing to feature frequent rain showers for the following three days after which time computer models were reported to have diverged from one another. Some predicted continuing rain showers into the weekend, whereas others forecasted a return to dry, warm weather. The reason for this is that weather systems are not periodic. That is, they do not return exactly to any state which previously existed. Of course there will always be times when it is sunny or it rains, but the exact configuration of the weather in terms of, say rainfall + wind-speed + wind direction + cloud depth (and so on) will never be the same at different time intervals. Similarly, our lives are not periodic. Although we may go each day to work, college, school or the local supermarket, it is never exactly the same experience, and in con-sequence we can never fully predict what is going to happen to us on any given day. And the main reason for this is that what hap-pens to us or the weather is sensitively dependent on initial con-ditions. This is an issue of great importance, but before it can be discussed further, it is first necessary to introduce the concept of non-linearity.

ASPECTS OF NON-LINEARITY

Viewing the universe as a linear phenomenon is commonplace in Western intellectual tradition. Within statistics it is routine to analyse data by way of linear procedures such as the regression analysis, where data are plotted on a straight line to depict the way two more variables will vary in a uniform way with respect to each other. In our own lives, too, it is normatively the case that we expect to live out a linear life involving employment, settling down with a long-term partner, increasing financial security and retirement into a peaceful stage of life free from the stress of competing for either success or merely to stay in work. Linear assumptions commonly abound within the social world, where any evidence to the contrary, such as the emergence of a radically new form of music and associated fashion, may elicit yet another cry that the end of society as we know it is just around the corner. We expect life to progress in a straightforward and uninterrupted way, and for centuries the natural sciences supported us in this belief.

One reason for this was that non-linear mathematical solutions to natural phenomenon are extremely complex. So, in consequence, linear techniques were used in analyses even though they did not reflect the actual conditions of the phenomenon being studied. As Stewart observed:

> Classical mathematics concentrated on linear equations for a sound pragmatic reason: it couldn't solve anything else. . . . So docile are linear equations that the classical mathematicians were willing to compromise their physics to get them. So the classical theory deals with *shallow* waves, *low*-amplitude vibrations, *small* temperature gradients.
>
> (Stewart 1990: 83; italics in the original)

And in the social sciences, there has been a general predominance of research that has followed linear statistical analyses, from cross-tabulations to complex procedures such as factor analysis.

In many cases the assumption of linearity does not cause great difficulty. Although, as shown throughout this book so far, many aspects of human existence are non-linear, this condition has not meant that findings from existing quantitative techniques such as surveys have not produced useful results. The linear analysis of the relationship between, say, social class and educational achievement, carried out by way of statistical techniques such as an analysis of variance, has produced useful findings for both social science researchers and policy-makers. Nevertheless, the use of such techniques ignores the fact that many natural and social phenomenon, even human consciousness itself, as argued above, are not linear. If this is so, then current forms of analysis will fail in certain circumstances to describe accurately the underlying trend of certain phenomena to display discontinuous change over time.

I am sure that the reader will by now understand the meaning of the term *non-linear* as used implicitly and explicitly throughout this book. It is clear from the preceding discussion that human evolution and individual development is a discontinuous process which, though bounded by species-level constraints, gives rise to extensive diversity. The progression of events over time is irregular. But, despite the overall accuracy of this statement, it still does not illustrate certain important characteristics of non-linear systems. To achieve this greater clarity, it is necessary first to consider the way non-linear equations are treated mathematically. Through this process, the reader will be able to gain important insights into the nature of non-linearity. However, do not be alarmed; in keeping with the rest of this book these ideas will be made accessible to most readers irrespective of their background in science and mathematics.

THE PROBLEM WITH NON-LINEAR EQUATIONS

As mentioned above, linear equations work well in many situations because they are easier to solve than non-linear equations. In consequence it is commonplace to linearise normally non-linear equations by discarding what may be considered difficult terms in the equations. Stewart (1990) provided a useful example of this process by describing the analysis of a pendulum. He describes the classical interpretation of the motion of a pendulum as the measurement of position x and velocity y. At rest the values x and y will be constant and the system will have the lowest possible energy. If energy is added to the system (you push the pendulum) then it will move from side to side, from left to right. When the pendulum is in the left sector (as it is viewed by the observer) x is positive, and when in the right sector x will be negative. Similarly, when the pendulum is at the bottom of the swing x will be zero. At the furthest point of the swing to the left, before the pendulum returns, the velocity y is zero. When the pendulum begins to return, when it swings once again back through the vertical position, y becomes positive and reaches a maximum value when it has passed through $x = 0$. If energy continues to be applied to the pendulum then at some point the pendulum will reach the vertical point above the ground (it would be pointing upwards). If energy were applied in such a way that both x and $y = 0$ then the pendulum could stay in the vertical position. However, as Stewart explains, this is an unstable position and even the slightest disturbance will cause the pendulum to fall. In consequence, the pendulum once again describes a downwards motion, accelerating as it does so. But, as Stewart points out, the dimension of time, which is clearly involved in the motion of a pendulum, has been left out of the equation. Although plotting position and velocity gives rise to an adequate qualitative description of the movement of a representative pendulum, it does not describe the full system. Consequently, from this model nothing is known about the length of time the pendulum spends in various states.

As a result, an equation such as $y = \pm\sqrt{(\text{constant} - 2\sin x)}$ (see Stewart 1990: 79) can be used to *plot* the movement of an idealised pendulum, but to solve the equation it is necessary to state the values of x and y at each value of t. If the action of friction is added to the problem then solving the resultant equation becomes extremely complex if not impossible. Instead, scientists interested in non-linear systems revert to a form of geometry known as topology:

> Topology is a kind of geometry, but a geometry in which lengths, angles, areas, shapes are infinitely mutable. A square can be continuously deformed into a circle, a circle into a triangle, a triangle into a parallelogram. All the geometrical shapes that we are taught so assiduously as children: to a topologist, they are one.
>
> (Stewart 1990: 63)

In other words, faced with an inability to solve non-linear equations, mathematicians have turned to the use of multi-dimensional space to plot the results of non-linear calculations. The most simple example of this is the logistic mapping.

THE LOGISTIC MAPPING

As has been demonstrated above, non-linear equations are not easy to solve. Instead of producing a single solution they most often give rise to many different solutions. One way of dealing with this is to use such equations iteratively, and plot the results. By so doing, important characteristics of the underlying laws of phenomena can be unfolded. The term *iterative* refers to the process whereby the results of a calculation are fed back into the equation and the calculation is repeated. This process, which can be repeated a large number times, can be illustrated by reference to an equation introduced by May (1976) as a means to model population dynamics. The equation (quoted in Ehlers 1992) has the form $x_n = r_x(1 - x)$; where x is a value between 0 and 1 which represents the level of the population, and $r = $ a constant. Used iteratively, the resultant x of each calculation is fed back into the equation to calculate the next value of x. Now this equation produces some interesting results. Following Ehlers's (1992) example, if $r = 1$ and $x = 0.2$ then after the first iteration $x_n = 0.2(1)(1 - 0.2) = 0.16$.

At all values of r below 3 the system eventually reaches a stable value, but when $r = 3$ something very interesting happens, the solution begins to oscillate between two values. Increase the value yet further and a further transition occurs such that the system then oscillates between four values. At approximately $r = 3.58$ the transitions describe a very large number of values, and at an r value of 3.58 a chaotic system forms such that an infinity of possible solutions arises. This general effect of increasing numbers of oscillating values is known as period doubling.

Of course, this is fairly abstract, so a practical example may help. Consider that the value x depicts the state of a social system which describes the amount of illicit drug-taking that takes place, and r represents the potential within the community for illicit drug-taking. Both of these measures will be historical in that their present values have arisen from events in the past. Assume that as the predisposition within the community to take illicit drugs is low, perhaps as a consequence of certain preconditions of use, both knowledge about and availability of such drugs are low (say, immediately after the 1939–45 war), the value of r will be low. Assume also a low starting point of x (there were few such illicit drugs used at that time). In consequence of these assumptions after a certain number of iterations, x will achieve a steady state. However, an increase in the value of r (the predisposition to use drugs present within the community) would give rise to a change in the prevalence of use within the community, and a new steady state would emerge at a higher level of use. However, if the community predisposition (r) rose to a high level (above 3 in the present example), then a steady state would not arise, instead levels of use would oscillate. At the value $r = 3.58$ the resultant oscillations would describe an infinite series (sometimes there would be almost total saturation, whilst at others there would be very little use). The most important aspect of this phenomenon is that the finding indicates that from one time period to the next it would be impossible to predict how much illicit drug use there would be within the community.

The usual way to represent this characteristic of a logistic mapping is to plot the results obtained from iterative calculations on a bifurcation diagram (Figure 1). The bifurcation diagram also

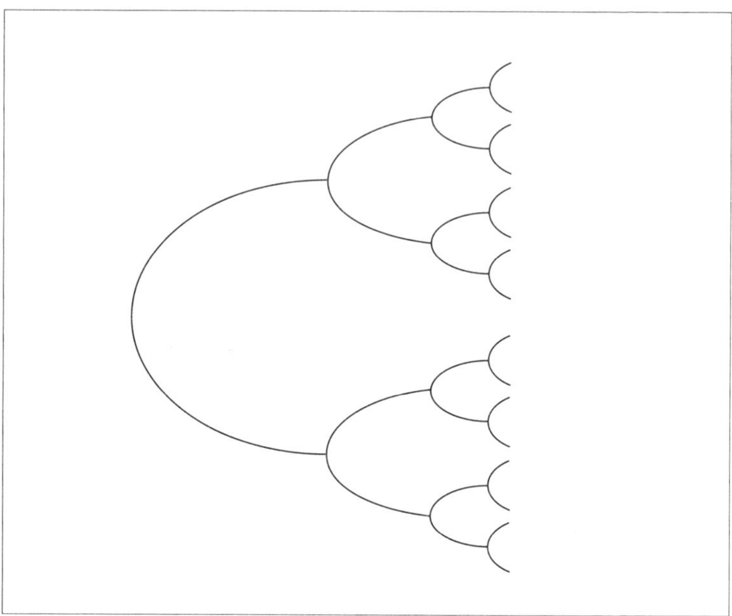

Figure 1 Bifurcation diagram: the vertical axis is value x and the horizontal axis is value r. As r increases to 3 and beyond, a period doubling is seen with respect to values of x. At $r = 3.58$ an infinite number of values arises. It is important to note that once period doubling has occurred, the outcome of subsequent iterations cannot be predicted. At the first bifurcation the outcome will vary between two solutions, at the second between four and so on until at $r = 3.58$ a very large number of solutions is possible, all of which cannot be predicted.

illustrates another important characteristic of chaotic systems: self-similarity. What this means is, if you took a portion of the bifurcation diagram after the point at which the system becomes chaotic and magnified it you would observe an exact replica of the original diagram, no matter what degree of magnification you selected. In fact there is an infinite replication of the original diagram. But the logistic mapping has even more unusual properties than this strange outcome. Mitchell Feigenbaum, a physicist working in Los Alamos in the 1970s, discovered a further important feature of the period doubling effect. In researching what happened at the point at which the system became chaotic, he first of all calculated the exact value of r at which each period doubling occurred. This was a long process and in trying to anticipate each subsequent number he noticed that the change from one value of r to the next followed an exact ratio; about 4.669. What he had discovered was that the ratio between successive values of r has the same value at each stage of the bifurcation diagram. A general formula from which the Feigenbaum number can be derived is given by Çambel (1993):

$$\delta = r_n - r_n - 1/r_n + 1 - r_n$$

where δ is the Fiegenbaum number (4.6692016090).

The importance of this finding is that it revealed the property of *scaling*; that is that within self-similar systems there is an exact copy of the whole reproduced time and again, becoming smaller and smaller. This means that any structural feature of the system appears repeatedly. As Hobbs observed:

All chaotic systems exhibit sensitivity to initial conditions; those arrived at by the period-doubling route also exhibit *self-similarity* across different scales. This means that any structural feature of the attractor appears repeatedly at different levels of scale, each a multiple of size larger or smaller than the one nearest it. Indeed, the ratio of the scales at which self-similar structures appear is related to the exponentially diminishing scales of each bifurcation during period doubling. Just as a pair of facing mirrors makes it possible to see an image of an image of an image of an image, ad infinitum, so this kind of attractor involves a set of curves or a collection of points within themselves an image of the set at a smaller scale, which in turn contains an image at an even smaller scale, etc.

(Hobbs 1991: 156; italics in the original)

This is obviously quite an unorthodox idea, as traditional geometric shapes do not maintain their structure when expanded. Imagine expanding a section of a sphere or a circle. You would discover that all these more usual shapes become straight lines when greatly magnified. Not so, however, structures which exhibit scaling. Visualising the magnification of a coastline provides the most common example of the scaling phenomenon. When viewed from orbit around the Earth the coastline of Britain appears irregular. There are many bays, estuaries and such which underlie this irregularity. If this view is magnified and instead of viewing, say, the whole of the Devon coastline you focus on only one bay, the irregularity remains the same. Clearly coastlines, and many other natural structures such as trees and river banks, are fundamentally different from orthodox geometric structures such as spheres and circles. The mathematician Benoît Mandelbrot coined the term *fractal* to depict these scaled irregularities which exist in the natural world. In terms of the discussion above, one can consider a bifurcation diagram to be a form of fractal presented in two dimensions.

So, fractals are scaled geometric shapes of infinite dimensions which depict natural phenomena such as clouds and coastlines, but what use is this knowledge to any study of human nature in general, and the problem of intoxication in particular? Well, the first point which needs to be noted is that a simple underlying law such as a logistic mapping can give rise to an extremely complex structure such as the bifurcation diagram. In addition, although the diagram is wholly determined by the logistic equation, the outcomes of its use in terms of the occurrence of a specific solution cannot be predicted. Once oscillations begin to take place the occurrence of a particular solution to the equation cannot be known in advance. In other words, the fact that a system is determined does not mean that outcomes can be known. The is an important difference between determination and certainty. But although non-linear systems are characterised by uncertainty, the degree of uncertainty depends upon the action of what are termed *attractors* within the system. Attractors are features of dynamic systems which affect the evolution of a system over time.

ATTRACTORS

Before we move on to consider attractors it is first necessary to introduce the idea of *phase space*. This is an important concept in that it describes the movement of dynamic systems within infinite space – a complex idea, but one which has been elegantly and simply defined by Gleick:

> Phase space gives a way of turning numbers into pictures, abstracting every bit of essential information from a system of moving parts, mechanical or fluid, and making a flexible road map to all its possibilities. Physicists already worked with two simpler kinds of 'attractors' [than strange attractors]: fixed points and limit cycles, representing behaviour that reached a steady state or repeated continuously. In phase space the com-

plete state of knowledge about a dynamical system at a single instant of time collapses to a point. That point *is* the dynamical system – at that instant. At the next instant, though, the system moves. The history of the system time can be charted by the moving point, tracing its orbit through phase space with the passage of time.

(Gleick 1988: 134)

What this means is that it is possible to depict a dynamical system as being at a given moment the resolution of all the elements which comprise the system (does this sound at all familiar?). This resolution can be denoted, diagrammatically, by a single point. For example, at the very moment you are reading this sentence (or perhaps more accurately still, a single word within the sentence) it is not without meaning to consider your existence at that moment to be an instantaneous resolution of all the components, biological, psychological, social, cultural and historical, which comprise your life. If it was possible to identify each and every component and depict this in an *n*-dimensional matrix, then what you are at one instant of time could be denoted by a single point within the matrix; that is you would have a unique *n*-dimensional address. But at the next time interval you will have moved to another point, and thus acquired a new *n*-dimensional address.

Now, at face value, it should be possible to move anywhere within the matrix which defines your life, but physicists and others working with non-linear systems have uncovered the fact that movement within phase space is determined by specific characteristics termed 'attractors'. That is, particular non-linear dynamical systems display differing patterns of movement; they are attracted to certain kinds of motion. This means that the point which describes the state of the system does not move at random, but instead describes specific forms of movement. The picture which is drawn to depict movement in phase space is termed a *phase portrait*. Now, typically, there are only four basic types of behaviour which occur in phase space and are depicted in phase portraits (Stewart 1990). In reality this is not quite true, as you will see, but it is far easier if these ideas are unfolded slowly.

The first kind of feature of a phase portrait is termed a *sink*. As the term suggests, this is where all points outside the single point which represents the sink move inwards; that is, they flow towards the sink. This describes a steady state, like a ball lying at the bottom of a bowl. If you place the ball on one of the internal sides of the bowl it will roll to the bottom and come to rest. A sink is, therefore, a point or steady-state attractor. The second feature is a _source_. As Stewart (1990) has described, a source is also a steady state, but one where adjacent points move away. Visualise a marble placed on top of a ball. If placed carefully it will stay there, but any slight disturbance and the marble will roll off the ball. A source is thus an unstable steady state. The third feature of phase portraits is the _saddle_. A saddle is a bit like sitting on a fence with one leg on one side and the other leg on the opposite side. You would be very stable with respect to movements forwards or backwards, but movements from side to side would be unstable. The pictorial representation of a saddle would show points moving inwards towards each other and then moving away at 90 degrees from their original direction of movement (in the example given of sitting on a fence, the lines moving inwards represent movement forwards and backwards, whereas the lines moving away represent side to side movement). There will be a point at the middle of the saddle (the balance point) where lines do not flow apart, and this will be a steady state. The fourth type is termed a _limit cycle_, which has both stable and unstable forms. Limit cycles are not single points like sinks, sources or saddles, they cover a region of phase space. Stable limit cycles are where adjacent points move inwards and become fixed in a periodic motion; they describe a cyclic motion. Conversely, unstable limit cycles, are regions where adjacent points move away.

Clearly these features of phase space are important as any point in the system which starts outside any one of them will move closer, for this reason these features are termed _attractors_. So, at this stage it would seem that, as Stewart observed:

for structurally stable systems in the plane – typical ones – the only attractors are

- single points
- stable limit cycles.

  If you like, the only long-term motions are

- stay at rest
- repeat some series of motions periodically.

  Or, more simply,

- sit still
- go round and round.

(Stewart 1990: 110)

There is one further form of attractor which, although not typical of general dynamical systems, is found in classical dynamics; this is the *quasiperiodic* attractor (Stewart 1990). Quasiperiodic attractors arise from the combination of several periodic attractors, each of which has an independent frequency. There are some fairground rides which act on this principle; the customer sits in a chair which rotates around the vertical axis, the rotating chair itself rotates upwards and over through the horizontal axis, all of which is attached to a large rotating wheel. Thus, in this example the unfortunate (or lucky) person is subject to three independent periodic rotations. If the period of the rotations has a shared measure, say one of 20 seconds and the other of 10 seconds then the resultant rotation is periodic with a period which is the sum of the two independent rotations (in this case, 30 seconds). However, if the independent rotations do not have a common measure, say 2 seconds and 32 seconds then the period will never repeat exactly, hence the use of the term 'quasiperiodicy' (see Stewart 1990 for a more detailed discussion of quasiperiodicy).

In terms of describing the movement of a point in phase space these attractors are fairly straightforward. If it is a steady state then it does nothing. If it is a periodic attractor it just goes round and round, and if it is a quasiperiodic attractor then the motion of the point which defines the phase space rotates in an integral series of periodic loops which describe a donut shape (loops within loops going round in a circle). But there is another kind of attractor which does not follow any of these patterns. This additional attractor does not stand still or repeat periodically but instead follows a complex trajectory which is unpredictable. Given its non-typical nature, this attractor came to be known as a *strange attractor*. Strange attractors are quite different from other types of attractor in the following ways:

> First strange attractors look strange: they are not smooth curves or surfaces but have 'non-integer dimension' – or, as Benoît Mandlebrot puts it, they are fractal objects. Next, and more importantly, the motion on a strange attractor has sensitive dependence on initial conditions. Finally, while strange attractors have only finite dimension, the time-frequency analysis reveals a continuum of frequencies.
>
> (Ruelle 1993: 64)

The concept of fractals was introduced above, but the other two features of strange attractors require some further explanation.

Taking the last feature first, what does it mean to say that strange attractors have a continuum of frequencies? This is not as complex as it may at first appear. First of all we need to understand the notion of a *mode*. Well, in simple terms a mode is a periodic motion around a mean position, effectively an oscillation or vibration (imagine a vibrating guitar string). In social science terms an example of a periodic oscillation could be an economic cycle (see Ruelle 1993) or, in terms of the present analysis, a weekly cycle of alcohol use. It can readily be appreciated that a periodic attractor has a single mode and a quasiperiodic attractor has a certain finite number of modes (the rotations within rotations – formally, a torus). The traditional way to think about this is to visualise what happens to water when you turn on the tap. When the tap is open only a little the water runs smoothly from the tap and appears to be stationary. When the tap is opened slightly more then regular oscillations of the water can be produced, thus displaying periodic motion. Opening the tap even further the pulsations become irregular, the more so the more the tap is opened (Ruelle 1993).

Now the traditional explanation of this was in terms of an increasing number of modes. The more energy applied to the system then the greater the number of modes of oscillation there would be within the water. Hence, dynamic systems become irregular through an increase in the number of oscillating modes. And, at some point, if energy continued to be applied to the water from a tap in the example above, then eventually the flow would become turbulent. There is no standard definition of turbulence, but we all know what it depicts: unpredictability and great irregularity. However, the mathematician David Ruelle and a colleague Floris Takens did not believe that transitions in the form of a dissipative flow (one where friction gives rise to strong interactions between modes) would lead to the generation of such multiples of modes, and thus form the obvious outcome: a quasiperiodic attractor. Of course, the first transition from a steady state to periodic motion, as described above in the example of water from a tap, is unproblematic. However, when more energy is added to the system a second mode forms and, through superimposition, this can give rise to a quasiperiodic attractor. Nevertheless this situation is unstable, as the small disturbances which arise from friction can break up this flow to give rise to one stable periodic flow and one unstable periodic flow. However, when three or more modes (or oscillators) form they do not give rise to a quasiperiodic attractor, instead they combine to create a strange attractor (Stewart 1990). Thus the formation of a strange attractor does not stem from a discrete increase in the number of modes as with a quasiperiodic attractor, but instead arises from the formation of a continuum of modes (i.e. a non-discrete number of modes).

Now, this may seem to be a fairly obscure point to make in a book on human nature and intoxication but this feature of strange attractors is central to the current analysis. Before this can be considered further, however, the concept of sensitive dependence on initial conditions needs to be introduced.

SENSITIVE DEPENDENCE ON INITIAL CONDITIONS

Sensitive dependence on initial conditions is perhaps the key condition for the existence of chaos in dynamic systems. What sensitive dependence refers to is the circumstance whereby the state of two (or more) systems will diverge if at time = 0 they have an asynchronous status. For example, you have two balls of exactly the same size, constructed of the same material and with the same mass, and you place one ball on the side of a mountain and release it, noting its path, and then repeat the exercise with a second identical ball. If there is only the smallest possible difference in the position in which the second ball was placed in relation to the first ball, then the balls will describe different paths down the mountain.

Perhaps the best-known example of sensitivity to initial conditions was provided by the work of Edward Lorenz who, at MIT, worked on calculations associated with convection (see Lorenz 1963). Convection occurs when two areas of liquids or gases with differing temperatures are placed together. Over time the dissimilar agents begin to circulate as the differing parts are first warmed then cooled. You can appreciate this phenomenon easily if you observe the action of water in a kettle beginning to boil. The same phenomenon occurs between two layers of air of differing temperatures. With a layer of cold air above a layer of warm air, convection currents will arise as a series of periodic whorls. Lorenz investigated this problem in the 1960s, and by reducing the problem to one of three dimensions (rather than the multi-dimensional nature of a real weather system) he found that the result did not result in a periodic or quasiperiodic attractor, but instead one which described a figure of eight, where each arm or whorl occupied a unique position with respect to all other spirals.

To visualise this, take a piece of paper and draw on it a figure of eight, then without stopping continue onto a second sheet, and then a third sheet and so on a large number of times. When you have drawn enough figures of eight place each sheet on top of the other. Now, recalling that each figure of eight is continuous with all the others, you will have drawn a three dimensional representation of a strange attractor. You will note that as each figure of eight is on a different sheet then, spatially, the representation never actually repeats itself. We know from the above discussion that periodic or quasiperiodic attractors repeat themselves (or almost nearly do so), but not so strange attractors. The second point to realise is that if the starting point of your reconstruction of a strange attractor (in three-dimensional space) began at a slightly different point, then an entirely different attractor would be produced. In reality, a strange attractor does not necessarily describe uniform figures of eight. The rotations around any arm of the attractor can take any value. So, in practice, to represent the form more accurately, sometimes you would have to draw several rotations round one arm of the figure before returning to the other, where you might draw only one.

If we consider this phenomenon in terms of the concept of phase space, then the reader can readily perceive that the outcome of a strange attractor is that the system under consideration will change status in an unpredictable way. As the point which represents the state of the system moves along the strange attractor, then the system will pass continuously through different states, none of which will be repetitions of any previous state. All that can be known is that in the next time interval the state of the system will change or stay the same. What Lorenz found was that with calculations of convection small changes in the initial parameters of the calculation produced a different solution, and that when plotted out the results depicted an apparently random set of trajectories around two lobes of a figure of eight:

> The trajectories of his equations, he realised, lived on something rather like a squashed pretzel. A surface that had two layers at the back, but merged to a single layer at the front. The point

that represented the state of the system would swing round one or other of these surfaces, pass through their junction, and then swing round again. Lorenz knew that trajectories of a differential equation *can't* merge. So what looked like a single sheet at the front must really be two sheets very close together. But that meant that each sheet at the back was double too; so there were four sheets at the back. . . . So four at the front, so eight at the back, so. . . . 'We conclude,' said Lorenz, 'that there is an infinite complex of surfaces, each extremely close to one or the other of two merging surfaces.'

(Stewart 1990: 139; italics in the original)

It was about ten years before Lorenz's work became widely known, but his findings now represent one of the earliest foundations of chaos theory. It is now fairly general knowledge that changes in the state of certain non-linear systems, such as the weather, cannot be predicted beyond a short period of time, and that this characteristic arises from the occurrence of strange attractors, which feature a continuum of modes or oscillators and sensitivity to initial conditions. Some chaos theorists, such as David Ruelle, hold the view that it is this last feature which is the best indicator of the presence of a strange attractor in a dynamical system.

CHAOS THEORY AND UNCERTAINTY

One of the most important aspects of chaos theory is that the consequences of the existence of strange attractors or the appearance of period doubling (the occurrence of bifurcations) means that specific outcomes cannot be predicted from existing states. Owing to sensitivity to initial conditions, a characteristic of all chaotic systems, the fact that a system may be determined by an underlying law, which may be simple and easily understood, does not mean that the progress of that system can be discerned. What it means is that we live in an uncertain world, one which will constantly give rise to the unexpected. As has been noted many times before, the 'butterfly effect', where small changes in one part of a system may produce large changes elsewhere in an unpredictable fashion, is increasingly being demonstrated to apply not just to simple experimental situations in the physical sciences, or with respect to weather systems, but to a wide range of systems in the biological and social world. Indeed, the scope of the applicability of chaos theory is only just beginning to be explored. Clearly, if the views of those researchers who claim that even events at the sub-atomic level can effect changes at the macro level can eventually be supported empirically, then the human sciences are on the verge of a major reinterpretation. As Hobbs has remarked:

> On the basis of the chaotic formalisms alone, no matter what bound one might specify for the system's final state, there is no neighborhood of the initial state small enough, even to the size of Planck's constant, to assure that starting within it will keep the final state within that bound. . . . it should be borne in mind that so long as the slightest perturbations in initial conditions grow exponentially with time, it would not take very long for changes even at the quantum level to become manifest at the macro level.
>
> (Hobbs 1991: 157)

Clearly this has profound implications for any human system within which it can be demonstrated empirically or theoretically that outcomes are sensitively dependent upon initial conditions. For, whenever this is found to be the case, then it can be hypothesised that the chaotic behaviour which arises will be mirrored at both macro and micro levels, relative to the current point of observation. For example, supposing you were undertaking a research study on the drinking patterns of a group of homeless men. They were selected initially on a number of criteria which included daily alcohol intake. The project proceeded well until at some point it was found that their drinking patterns began to diverge from one another rapidly and unpredictably. Now, you may have supposed that there was something wrong with your original sample and the subjects were different from each other in a way you had not anticipated. So you would draw up another sample and continue with your study. But now, having encountered chaos theory and the phenomenon of the exponential divergence of systems with differing initial conditions, you may be drawn to conclude that you were witnessing period doubling or the action of a strange attractor. This would not be a comforting thought because, given sensitivity to initial conditions (which may lie at the cellular or even quantum levels), the reason for this divergence may be forever beyond your ability to record. In fact it may be that empirical understanding of human nature will always be severely limited. However, it is still early days in the formation of this new area of research, and there is still a great deal which can be done to move these ideas forward both empirically and theoretically. And the ways in which social scientists are beginning to work with chaos theory is the subject of the next chapter.

Chapter 8

Working with chaos

The previous chapter provided a detailed discussion of the basic principles of chaos theory and complexity. Attention was drawn to some of the essential features of chaos, particularly those relating to bifurcations (or period doubling), self-similarity, strange attractors and sensitivity to initial conditions. From this, it was established that dynamic systems which display sensitive dependence on initial conditions have a time evolution which is unpredictable. The state of a system at time + 1 cannot be known unless all the conditions at time = 0 can be specified. Clearly, in practice, this is an impossible task because, as discussed by Hobbs (1991), over time even differences at the quantum level will eventually be expressed at the macro level. It would not be surprising, then, if the reader concluded that chaos theory takes empirical research beyond the bounds of scientific certainty into the realms of the unknown and unknowable. However, this need not necessarily be the case. Of course the implications of chaos theory suggest that we will be bound by limitations in our endeavours to explain the social world and individual action, but this does not mean that nothing useful can be discerned. There may be limitations, but, as with weather forecasts which do in part predict the future, the application of chaos theory to the social sciences may also lead to new forms of understanding. At the very least we will know why we can predict so little.

Arguably, sensitive dependence on initial conditions is the key condition for the existence of chaos in dynamic systems. As the reader will recall, sensitive dependence refers to the circumstances whereby the state of two (or more) systems will diverge if at time = 0 they have asynchronous status. As discussed in chapter seven with respect to a hypothetical group of homeless people with alcohol problems, individuals who are close at time $= 0$ may be arbitrarily apart at time $+1$. Essentially, any arbitrarily small differences which exist will become amplified over time. You will recall that Lorenz's work on weather patterns first established this condition of certain non-linear systems through the iterative use of convection equations. What he found was an immanent exponential divergence within the system which gave rise to what came to be known as a strange attractor. As you will recall, strange attractors have a trajectory which over time spirals back in to itself without ever completely repeating any previous cycle. In effect, it was found that the movement on a strange attractor contracts in some directions and stretches in others. It is this folding and stretching which gives rise to the fractal structure which has been found to be characteristic of chaotic motion (Brown 1996a).

In consequence of these findings it has been argued that one of the first problems to be encountered with research which does not acknowledge underlying chaotic structures concerns the limitations of linear statistical analyses. Due to the existence of sensitive dependence on initial conditions, traditional statistical methods do not work with chaotic systems. As the reader will readily appreciate, any small discrepancy in a quantitative measurement will ultimately lead to large errors. In addition, and for the same reason, the replication of experiments and the use of longitudinal studies also become problematic. This is not to say, though, that no prediction is possible, but only that in such systems predictions are very limited. As we know, in areas of constantly changing weather such as the British Isles, weather forecasts can be accurate, but only over a few days.

The way such limited accuracy is achieved is, very crudely, through the averaging of different trajectories (on a strange attractor) which are initially close to each other (Brown 1996a). For example, although it is very difficult to make predictions about single individuals, through the statistical aggregation of collective data it can be possible to predict what groups of individuals may do. This suggests that if predictions about the path of chaotic systems can be made, even if over only short periods of time, then there must some order within chaos, it is just that it does not last long.

In general, the use of standard linear procedures featured in traditional statistical analyses may fail to reveal important trends which can exist in certain non-linear systems: 'Poor analytical results (e.g., low R_2 values and a lack of statistical significance) are to be expected when analyzing chaotic systems with standard statistical methods' (Gregersen and Sailer 1993: 793). This limitation arises because in systems which exhibit a chaotic structure, states or values which are initially close can separate exponentially fast. Usually, standard social science analyses are not able to cope with such divergence, as analyses are generally based upon a search for similarities. In consequence, outliers – data points which do not fit standard statistical models – are usually eliminated from the data set. When large-scale divergences are found, then it is customary practice either to endeavour to standardise through selecting new sub-samples, or to seek to discover the 'missing' variables which, if included in the analysis, would enable the divergence to be explained and eliminated. However, when chaos is present within the system under investigation, then divergence would be the norm. As Gregersen and Sailer have observed, if states relevant to the analysis are:

> near the boundary between diverging and non-diverging parts of a chaotic system, the accuracy of such [linear] statistic models could be spuriously low. Even if hundreds of very similar entities all diverge, another hundred with nearly identical profiles might not, since the underlying causal laws themselves produce discontinuous behaviour.
>
> (Gregersen and Sailer 1993: 793)

So, traditional analyses will not find statistical significance if the underlying structure of the phenomenon is chaotic. Clearly, we need new ways of looking at certain social, cultural and psychological issues. First of all, we need new analytic techniques which will be able to detect the underlying structure of social events and thus reveal the existence of any chaotic behaviour. To do this, a form of modelling chaos which allows for the special difficulties encountered in the social sciences when compared with mathematics or physics, needs to be developed. Although such work is still in progress, current theoretical and empirical developments in this area are providing new insights into the nature of non-linear and chaotic social phenomenon.

MODELLING CHAOS

Of course the easiest way to detect chaos would be to demonstrate the existence of sensitive dependence on initial conditions.

Once demonstrated, you could then move on to design an appropriate analytic technique to measure the effects, certain in your mind that chaos existed. In essence, you would then need techniques with which to test the behaviour of the system under investigation.

To test for the existence of sensitive dependence on initial conditions it may be necessary to set up a mathematical model of the system under investigation, as was done in chapter seven with respect to alcohol use in a idealised community. Once this has been done, the model can be run iteratively with differing initial conditions. After a number of iterations the model will settle down to some final configuration. Note this configuration and then repeat the trial with a different set of initial parameters. This process can be repeated a number of times, each time using a different set of initial conditions. If the final states arrived at settle down to similar configurations then the system that has been modelled is stable. However, if the final states of different trials are widely different from each other, then the system can be considered to be chaotic (Saperstein 1996). You will recall that this is the process followed by Lorenz (1963), whose work with convection systems led to the development of an understanding of the chaotic nature of weather systems.

In developing such models it is not necessary that the whole system be understood. Indeed, as Saperstein (1996) has noted, if a working knowledge of the whole system existed, then questions about its underlying structure would be irrelevant. In practice, it is possible to gain meaningful results with models that are much reduced in complexity. For example, as mentioned above, Lorenz (1963) reduced the problem of convection within a system comprising infinite dimensions to one where there were only three. Nevertheless, despite this large simplification, his model produced important insights into the structure of weather systems. In general the evidence is that chaos found within simplified models is generally unproblematic with respect to conjecture about the actual system. As Saperstein has argued, there is no evidence of chaotic regions in simplified models disappearing when new dimensions or variables are added. Indeed, he states that:

> Theoretical experience with specific mathematical models of real phenomena (the Navier-Stokes equation for fluid flows; the recursive equations' modelling of the evolution of tripolar

systems from bipolar ones [Saperstein 1991] . . .) suggests that the regions of stability (areas of the absence of chaos) decrease in extent when additional variables come into play.

(Saperstein 1996: 150)

In practice, complete models do not yet exist with respect to social, psychological or economic systems, hence in developing new ways of analysing chaos there must be reliance on analogies from the natural sciences, where complete models do exist: 'Empirical experience with fluids indicates that chaos (turbulence) appears earlier and stronger when new variables, such as temperature differences and heat flows, become important in the system' (Saperstein 1996: 150). However, one must be cautious of the significance of *stability* in simplified models. For the reasons already given, once a simplified model is extended to include new variables, then chaotic regions previously absent or undetected may appear.

Hence, when looking for evidence of chaos within social and cultural phenomenon, it is not necessary to have a full description of the system being researched. It is permissible methodologically to work with simplified models. However, even though this may be the case, there are still considerable problems in trying to use the approach within the social sciences. Although there has been some success in modelling political and economic behaviour, any significant success in modelling behaviour in other areas in the field of social science, particularly with regard to individual action, has yet to be recorded.

Gregersen and Sailer (1993) have looked at this topic with respect to a simple model of interaction between commercial marketing, production and effectiveness. They argued that in principle once a starting value for each of these three variables has been defined, it would be possible theoretically to monitor changes in values over time to see whether stable states were achieved, or whether, conversely, unpredictable changes would be found to arise. However, they noted that to undertake such analyses it would first be necessary to establish the values of the variables which comprise the idealised model, and that to do this would require the collection of quantities of data which have so far proved beyond such empirical studies. Hence, in light of these problems, they claimed that, in agreement with the views expressed above, an acceptable approach would be to construct a simplified mathematical model which would represent central aspects of the phenomenon under investigation. The simple logistic equation cited in the previous chapter is an example of such an approach. Gregersen and Sailer (1993) followed this tradition in establishing through experimentation a quadratic equation which depicted a two-person social system. They then estimated the values of terms within their experimentally derived equation and computed the results iteratively to illustrate the dynamics of the system under investigation. In terms of supporting the general nature of their model, they noted that their equations, held to represent a two-person social system, were of the same class as Mandelbrot's fractal generating equation (see chapter seven).

As Gregersen and Sailer were unable to fit their model to actual data sets, they were, in practice, working with *meta-models*, as they record. This is a legitimate approach when the collection of data proves difficult (see Stewart 1990). By these means they found evidence of discontinuous divergence and non-divergence within their two-person system; that is, the stretching and folding which characterises fractal structures. Despite the fact that their work did not incorporate empirical data, the importance of this work in demonstrating the possibilities of such studies within the social sciences should not be underestimated:

> We realise that this is, in fact, armchair modelling; however, we insist that there are some very important lessons to be gained from the fact that these simple models produce such unusual patterns of behaviour. One reason why these lessons are important is that chaos seems to be quite common in social contexts.
>
> (Gregersen and Sailer 1993: 791)

THE LYAPUNOV EXPONENT

Although it may be currently difficult (or even impossible) to provide comprehensive models of social phenomena due to problems of data collection, there are nevertheless alternative approaches which can provide significant insights into the underlying structure of very intricate systems. One such approach involves the calculation of *Lyapunov exponents*. Lyapunov exponents are a measure of the mean rate at which nearby points in dynamical systems move with respect to each other, and in one-dimensional systems there are three variations:

- the Lyapunov exponent is < 0 (the system is stable and periodic),
- the Lyapunov is $= 0$ (the system is marginally stable),
- the Lyapunov is > 0 (the system is chaotic) (see Brown 1996a).

To simplify this for a moment, imagine you have a smooth bowl and you release two balls, one on each side of the bowl. You would find that both balls move down the side of the bowl and come to a steady state at the bottom. Similarly, if you targeted two identical satellites at the moon and launched them into space with the same force and trajectory, then, if the amount of energy used was calculated correctly they would both achieve identical orbits around the moon, being separated only by the time delay between each launch (that is, they would achieve the same periodic motion). If, however, instead of there being only one moon there were instead a very large number of moons all rotating around each other, then unless the initial trajectories were absolutely identical in all aspects, including time of launch, different trajectories would quickly evolve for each satellite; that is, there would be aperiodic motion. As you will have appreciated from earlier discussion, with large numbers of bodies in orbit, the orbit of each satellite would be sensitively dependent on initial conditions. Now, if you did not know how many moons there were in orbit in the area in which you were to launch the satellites, then you could not predict the outcome for each satellite in terms of resultant orbits. However, if you performed a series of analyses which compared changes in the relative trajectories of the satellites over time, or, in other words, you calculated accumulated errors in predicting one orbit from the other, you would then be able to discern if there arose: (a) a steady single point attractor, (b) a periodic attractor or (c) an aperiodic, strange attractor (for simplicity, the possibility of a quasiperiodic torus has been put to one side). In very simple terms, this is how a Lyapunov exponent is calculated.

Lyapunov exponents are generally held to be related to other measures of chaos:

Ruelle (1983, 1989) suggests that K entropy [a measurement of the loss of information over time characteristic of chaotic systems (in general, entropy is a measurement of increasing disorder in a system)] equals the sum of the positive [Lyapunov exponents], implying that positive entropy exists in the presence of chaos. The [Lyapunov exponent] is linked to the information lost and gained during chaotic episodes (Ruelle 1980; Shaw

1981, respectively), and is closely linked to the amount of information available for prediction.

(Brown 1996a: 57)

In consequence, if it is possible to calculate a Lyapunov exponent for a system then the resultant value would describe the structure of the system; whether it was stable, partly stable or chaotic.

In practice, calculating Lyapunov exponents is complex. However, as Brown (1996a) notes, there are two algorithms currently available; the Wolf algorithm (1986) and the Eckmann–Ruelle algorithm (1985). The Wolf algorithm follows the movement of two nearby points on the attractor to measure changes over time. According to Brown (1996a), who has provided a broad discussion of the derivation of the Lyapunov exponent and the Wolf and Eckmann–Ruelle algorithm, the Eckmann–Ruelle algorithm provides advantages over Wolf by calculating the complete range of Lyapunov values in a system, whereas recovering more than one with Wolf is reported to be extremely difficult.

Saperstein (1996) has used the Lyapunov exponent to review the underlying structure of international arms competition. Saperstein looked at three political theory questions: (a) Are bipolar international systems more or less stable than corresponding tripolar systems? (b) Is a system of democratic nations more or less stable than corresponding systems of autocratic states? and (c) Is a system of nations that strives for a balance of power via shifting coalitions of states more or less stable than one in which each nation individually seeks to balance the power of others? Of these, only the first will be reviewed, though the interested reader is strongly recommended to refer to the original text for further discussion of this valuable work.

The variable Saperstein selected to describe national behaviour with respect to a predisposition towards war was a measure of 'devotion' to war preparation. This, he determined, could be given as the ratio of the annual expenditure on arms to the gross national product in the same year. The value of this variable must lie between zero and one. The bipolar world was then modelled by the following equation:

$$X_{n+1} = 4_a Y_n (1 - Y_n)$$

$$Y_{n+1} = 4_b X_n (1 - X_n)$$

where X_n = the devotion of nation X to war in year n.

The purchase of arms in year $n+1$ is assumed to be a proportional response to the amount purchased by nation Y in the previous year. The non-linear term $(1 - Y_n)$ is a limiting factor relating to the nation's GNP. When computed, it is possible to plot a curve which represents the critical relationship between the proportionality parameters a and b (recall that expenditure in one nation is proportional to that in the opposing nation in the previous year – hence the proportionality parameters). The region above the curve, in which the two Lyapunov exponents are positive, is the model's chaotic region. Now, when the model is extended to three nations, the effect of increasing that third nation's proportionality parameter from zero is to decrease the area of stability depicted from the resultant curve. In other words, the model illustrates that as a third nation in a tripolar system increases its expenditure on arms from zero the system becomes increasing unstable. Hence, as Saperstein concludes, a tripolar world is less safe than a bipolar world.

These early attempts at modelling social systems, which also include significant new work in economics and approaches such as spectral analysis (see McBurnett 1996), are clearly important as they mark the beginnings of a new rigour in understanding social events. Quite rightly, in some respects, social scientists have been criticised for a lack of thoroughness in their investigations. Too often research resembles works of literature rather than works of science. It is common to encounter research reports where the author's values appear to dominate, or where findings are comprised merely of abstracted statements about data derived from the ubiquitous cross-tabulation. A move towards modelling the social world mathematically may lead to new insights into the nature of underlying structures, and in this regard attempts to use the techniques and insights of chaos theory are to be welcomed.

However, this welcome should not be blind. As insightful as the above examples of new work are, they are also subject to limitations, not least because of the enormous difficulty there is in collecting data sets large enough to use with mathematical models. This being so, it is also important to look at other ways in which the social world can be approximated, approaches which are able to estimate more closely the dynamic nature of such systems.

BOOLEAN NETWORKS

One method of rigorously studying patterns of change and stability generated by the non-linear structure of dynamical systems is by way of Boolean networks. Boolean networks are systems of binary elements which are activated through coupling by other binary elements in the system. Relationships between elements are governed by a switching function which defines a response in terms of adjacent elements. For example, if element A is set to 1, element B to 0 and element C to 1, and the relationship between them is determined by an *or* function with respect to A and $B + C$, and an *and* function for B, C and A then the initial configuration 1 0 1 in time t will switch to 0 0 1 in $t + 1$, and 0 0 0 in time $t + 2$. For networks of increasing size it can be readily appreciated that a change in any part of the system, either in terms of a random change in the state of an element (1 or 0) or in the quality of a function (say, an *or* function shifting to an *and* function) may effect the whole system in t_x.

According to Kauffman (1993) the utilisation of switching Boolean networks is central to the study of complex systems because:

- Boolean networks attune or synchronise the action of many thousands of elements in a system to produce ordered, complex or chaotic behaviour;
- the binary Boolean network is an extremely precise way to demonstrate non-linear behaviour in dynamic systems;
- given that Boolean networks have definable properties (outcomes from particular parameters can be determined) then use of such networks can lead to the generation of a new statistical mechanics – that is, the occurrence of specific properties will have a specific statistical distribution;
- despite the simplicity of the underlying rules which govern Boolean networks, three different types of behaviour arise; ordered, complex and chaotic;
- given the above, Boolean networks provide the means of exploring the promulgation of order, complexity and chaos in dynamical systems.

Kauffman (1993) has produced an excellent review of the main features of Boolean networks, which the reader is strongly encouraged to refer to for an up-to-date appraisal of this approach to understanding complex and chaotic systems. The following is merely a summary of those aspects of such networks which provide further insights into the concept of chaos within the psychological, social and cultural domains.

Kauffman (1993) summarised the behaviour of such networks in terms of the number of elements in the net (N), the average number of inputs to each element (K), and the biases on the sub-set of possible functions which feature in the net (P) (i.e. the distribution of particular functions, which may change over time). Having done so, he then analysed the routine behaviour of networks in terms of fixed values of N, K and P. What he did was to assign at random specific values of N and K and then sample the resultant networks to determine what forms of behaviour arose with what frequency of occurrence. From this, he was able to discern particular cases of the relationship between K and N which gave rise to differing outcomes at the network level. More specifically, he was able to summarise the cases where $K = N$, $K = 2$ and $K = 1$. From each of these boundary cases the following conditions arise:

1 $K = N$ gives rise to the largest possible cluster of Boolean networks defined by N elements and the maximum of amount of disorder. The length of the attractor within the system expands rapidly with increasing values of N. When $N = 200$: 'at a microsecond per state transition, it would require about a billion times the age of the universe to travel the attractor' (Kauffman 1993: 194). In addition, networks where $K = N$ show maximum sensitivity to initial conditions. However, despite the existence of chaotic behaviour, the number of attractors is relatively small. As Kauffman notes, a system with 200 elements would have only 74 alternate patterns of behaviour. However, due to inherent instability any small perturbations can switch the network to a different attractor. Between $K = N$ and $K = 2$, random Boolean networks exhibit chaotic behaviour. Although state cycles are seen to progressively lengthen as the value of N increases, the network still shows sensitive dependence on initial conditions. However, as K approaches 2 the properties of the chaotic region change radically.

2 When $K = 2$, order emerges within the Boolean network. This is in some ways one of the most exciting aspects of Boolean networks. Networks which are random except for the condition that the number of inputs to an element cannot rise above 2 show high levels of order. Citing some recent work by Derrida and

Pomeau (1986) Kauffman demonstrates that where $K = 2$ the overlap between two initial states as compared to their subsequent states decreases over time. That is, when $K = 2$, states with differing starting conditions converge over time. They do not merge totally, but reach a fixed fraction of overlap, the extent of which depends on K. This does not occur when $K > 2$. So, as K reduces in value from N to 2, order appears in previously chaotic systems! This is a remarkable transition, and one which has also been noted by Ruelle, but in a different context: 'for sensitive dependence on initial conditions to occur, *at least three oscillators are necessary*. In addition, *the more oscillators there are, and the more coupling there is between them, the more likely you are to see chaos*' (1993: 81; italics in the original). Hence, if there are fewer than three oscillators ordered motion will be expected.

3 When $K = 1$ the network forms separate loops which are independent of each other. Hence a modular structure of isolated subsystems forms, the product of which gives rise to the behaviour of the whole system. As N increases, lengths of state cycles increase slowly, whereas the number of attractors increases exponentially (Kauffman 1993).

What arises from this overview of specific initial states of Boolean networks is that in important respects they can be used to demonstrate in a controlled way central characteristics of dynamical systems, particularly the conditions from which chaos can be seen to arise. In this respect, Boolean networks can be used as a tool to explore phenomena found in empirical studies in both the natural and social sciences. However, Kauffman's work also draws attention to a moment in the evolution of dynamical systems which has so far been overlooked. That is, what it is that happens at the point of transition between chaos and order?

The basic question here, and one that Stuart Kauffman has asked, is what properties are represented in networks where $K = 2$ such that order arises? The answer is that these networks develop areas where elements are frozen in one of the states 1 or 0. Such areas remain in the fixed state irrespective of events outside the fixed area. Kauffman (1993) has referred to these areas as *frozen cores* which create *percolating* walls of constancy which, in turn, divide the network into distinct areas, each cut off functionally from any others. The boundary region between areas which are freezing and percolating is of great interest, as this is the region where chaos becomes ordered and order becomes chaotic

(Kauffman 1993). He notes that there are two ways of such order being established: through *forcing structures* and *internal homogeneity clusters*.

A forcing structure is a structure which is based upon a system of Boolean functions whereby the status of an element is determined by the status of one of its inputs (irrespective of the status of any other inputs). For example, if an element with the value 0 has three inputs each with the Boolean function 'Or', then a change in any input will change the element no matter what the status was of any other element:

> I define as a *canalyzing Boolean function* any Boolean function having the property that it has at least one input having at least one value (1 or 0) which suffices to guarantee that the regulated element assumes a specific value (1 or 0). 'Or' is such a function. So is 'And' since, if either the first or the second input is 0, the regulated locus is guaranteed to be zero at the next moment.
>
> (Kauffman 1993: 203)

The consequence of this can be appreciated if you visualise a number of elements in a Boolean network with three inputs, all of which are regulated by the 'Or' function. At time t all elements have the value 0. If at $t + 1$ one element takes the value 1, then this change will be forced throughout the sub-network governed by the 'OR' function. The forced transition will be halted when a function other than a canalysing Boolean function is encountered, such as the 'Exclusive Or' function. Kauffman terms such circuits *forcing loops* or *descendent forcing structures*.

The second way regions of order can emerge from chaos is by way of *internal homogeneity clusters*. The notion of biases in the distribution of functions within a network has been mentioned above. It was noted that P was given as the biases on the sub-set of possible functions which feature in the net (i.e. the distribution of particular functions, which may change over time). Consider for a moment a situation where there are x number of input variables where 1 or 0 may occur all the time, or only part of the time; in other words the probability of 1 or 0 is between 0.5 and 1.0. Now, if P is given as the larger proportion of the 2^K positions in the network which are either 1 or 0, then P ranges from 0.5 to 1.0. Consequently, P is thus a measure of the internal homogeneity of a Boolean function

(Kauffman 1993). It is a measure of the proportion of the combinations of actions which give rise to a response of either 1 or 0.

Work by Derrida and Stauffer (1986), Weisbuch and Stauffer (1987) and de Arcangelis (1987), summarised by Kauffman (1993), found that if P is larger than a value P_C, then there are formed within the network regions of elements frozen at the value 1 which are surrounded by islands of connected points, isolated from each other by the frozen region, which fluctuate from 0 to 1 and back to 0. However, when P is close to 0.5 this does not happen. Instead, small islands of frozen values form which are surrounded by a percolating web of elements which switch between 1 or 0 by way of complex cycles. Hence, the value P_C describes a point of phase transition in a complex dynamical system. The actual value P_C takes is dependent on the structure of the network. As Kauffman (1993) records, for a square lattice where $K = 4$, $P_C = 0.72$.

The consequences of this are that in the absence of descendent forcing structures or where $P < P_C$ then dynamical systems have very large attractors which are sensitive to initial conditions. However, in the presence of forcing structures or loops or where $P > P_C$ then percolating frozen islands or regions form which are stable and thus not sensitive to initial conditions. The fact that within these networks order can be seen to arise out of chaos is centrally important:

> [results show] that a *phase boundary* separates networks that exhibit frozen, orderly dynamics from those that exhibit chaotic dynamics. The existence of this boundary leads us to the very general and potentially very important hypothesis: Parallel-processing systems lying in this interface region between order and chaos may be those best able to adapt and evolve. Further, *natural selection* may be the force which pulls complex adaptive systems into this boundary region. If so, we begin to have a powerful tool with which to examine the collaborative interaction between self-organization and selection.
>
> (Kauffman 1993: 218)

This is an exciting finding because it gives an important insight into the way an adaptive potential may reside within complex natural systems. To demonstrate the mechanism through which such adaptation could take place through natural selection, Kauffman summarised the work of Ashby (1960).

Ashby produced a simple but entirely convincing model of how natural selection could lead to the design of a brain. His idea was based upon the premiss that although there could be a large number of variables within the system from which a brain could arise, only a sub-set of these would be essential. If these 'essential' variables did not appear in the final structure of the brain then it would not work. Now, he argued, first of all through the process of evolution a 'brain' could have any of the existing variables present within the founding system, and, second, over the course of time the brain system would settle down to an attractor with a finite number of variables. Now, if the essential variables were present, then the system did nothing. However, if some of the essential variables were missing then a *jump change* in one of the parameters of the system would occur. As we know from the above discussion of Boolean networks, if a parameter is changed then the system will reconfigure and eventually settle down to a new attractor. In the case of designing a brain, it was argued, such jump changes will recur until the system settles down to an attractor which includes all the essential variables for the selected brain – a very simple idea, but one which proved to be extremely effective. These ideas led Ashby to be able to design a autopilot, called a homeostat, which could fly an aircraft in straight and level flight.

As Kauffman points out, Ashby's model draws attention to the evolutionary and selective advantage of working with systems at the borderline of order and chaos. If a system is too ordered then jump changes to new configurations would be difficult; if too chaotic, then ordered structures would not form. As Kauffman contends, it would seem that these findings suggest that evolution has led to the propagation of natural systems which stand at the edge of chaos: 'Thus we are led to a bold hypothesis: Living systems exist in the solid regime near the edge of chaos, and natural selection achieves and sustains such a poised state' (Kauffman 1993: 232).

Clearly, Boolean networks provide an important means through which central characteristics of complex and chaotic systems can be explored. Kauffman's work is directed towards the study of adaptation in biological systems, but there is no reason why this approach could not be used to examine psychological, social or cultural systems. As has been discussed at length above, there is good evidence for concluding that adaptation plays an important part in affecting outcomes both at the individual, cognitive level, and with respect to social action such as that involved in the laws of social exchange. Hence, through controlling for specific Boolean functions and their distribution, and the size and values of K and P, it may prove to be possible to utilise Boolean networks to model observed psychological, social or cultural events, and thus reveal important aspects of the underlying dynamics of such networks or systems.

Similar approaches have already been adopted in the study of political behaviour by way of cellular automata, a more simplified form of dynamic network based upon a two-dimensional lattice where each element will have one of eight possible states. Within these networks functions are used to determine the state of an element in conjunction with the states of the adjacent four elements (Kauffman 1993). Cellular automata have been used to explore the dynamics of political systems such as voting patterns. Brown (1996b), who provides an overview of some current research in this area, has argued for the importance of such simulations in that, unlike other approaches such as the Lyapunov exponent or spectral analysis which are based upon discrete depictions of chaos, they can be used to represent the spatial and temporal changes which characterise dynamical systems. They can illustrate all the main features of such systems: steady states, periodic cycles and chaos.

Overall, the contribution that the study of Boolean networks and cellular automata may bring to the study of complex social and psychological systems must not be under-estimated. This is a new science, and these techniques are only just beginning to be exploited by social scientists. As the above discussion has shown, through the use of such networks it is possible to understand the underlying dynamics which give rise to static, ordered or chaotic systems. But this approach also offers the possibility of viewing systemic changes over time as qualitative changes in the way the system is presented. Boolean networks and cellular automata both have the characteristic of changing the value of constituent elements over time. As you know, for Boolean networks this means that each element will take the value of 1 or 0. Now, if these changes were represented qualitatively by changes in, say, colour on a graphical display, then the researcher would have the added advantage of being able to study patterns of change in real time as they occurred with the simulation. Programs which exploit this ability to represent complex changes visually already exist and are utilised in a variety of differing ways, from exploring the behaviour of slime moulds to uncovering interactions in flows of traffic. One of the more advanced programs is StarLogo, produced by the Epistemology and Learning Group at the MIT Media Laboratory. However, there are many others available via the World Wide Web. What each of them provides to varying degrees is the ability to view the continuing patterns of spatial and temporal changes which take place in such systems. Consequently, developing a qualitative component to Boolean and other related simulations may offer new insights into the unfolding of chaotic and complex systems.

PATTERNING CHAOS

Traditionally, one of the main problems with evaluating underlying structures in social systems is that, as with traditional statistical techniques, what is being looked at is a snapshot of the system made at one particular moment in time. A mathematical model of a system may disclose that the data have a fractal structure, or the calculation of Lyapunov exponent may reveal the existence of chaos, but neither approach offers a great deal in terms of providing insights into the range of activities which may be possible. In other words, you will know that the evolution of the system cannot be predicted, but you will know little about what that means in qualitative terms. For example, if it were possible to model a person's use of alcohol such that you could determine that the underlying structure was chaotic, or if you could do this for a whole community (say, the population of a village), these findings would say nothing about what *kind* of behaviour would occur, only that patterns of use were unpredictable. This general context is important because the existence of chaotic structures does not mean that everything and anything is possible, only that what is possible may occur at any time. This can be deduced from two separate findings. First, although a strange attractor occurs in infinite space (there are an infinity of dimensions) it has only a finite dimension (Ruelle 1993). Second, earlier discussions regarding adaptation have drawn attention to the existence of species-level constraints, which not only apply to evolutionary events but also to what it is possible for human beings to execute behaviourally. Now, at face value, this does not seem to tell us very much. We may know what is possible but, as we cannot tell what is going to happen next, it does not seem to help us much. Well, this is only partly true. Although chaotic systems are unpredictable, they are still deterministic. Whatever happens is determined by some underlying structure. So, if we can generate a simulation which operates on the same deterministic basis as the real world systems we are interested in (say, voting behaviour or drinking patterns), then by reviewing qualitatively what happens by way of graphical representations of the evolution of the system we may gain new insights.

Well, you may be asking at this point how viewing patterns is going to assist discovery in a scientific enterprise. The answer to this lies in chapter five and the discussion of reason, rationality and individual action. You will recall that through a discussion of the work of Edelman and Cosmides it was argued that people have different levels of cognition from which their actions stem. These were given as reason, adaptive rationality and formal, scientific rationality. In certain situations which relate to events we have evolved to encounter, the evidence is that we operate on the basis of our adaptive rationality, rather than through the execution of formal logic. Although these ideas will be familiar to you, with respect to cognitive facilities there is at least one further important outcome of this finding. If, as has been argued, an adaptive category of cognition has arisen by way of natural selection, then it can be expected that the recognition of patterns would feature in the resultant form of adaptive reasoning. (What is meant by the phrase 'one picture is worth a thousand words' if not that visual representations of complex events are more easily assimilated than long passages of text or mathematical formulae?) If this is so, then it is likely that we have an innate ability to discern important information from visual representations of chaotic social systems.

Is there any evidence for such a claim? Loye (1995) has produced an interesting paper which provides evidence that we can in fact make predictions about non-linear events, and that an ability to see patterns in chaotic systems is a part of this ability. In this paper Loye cites two studies which have looked at the ability to predict: the work of McGregor (1938) and of Cantril (1938).

In 1936 McGregor carried out a study in which he asked 400 students at Dartmouth, Bennington and Columbia colleges, and teachers at MIT to predict the outcomes of several current events. These events included Franklin Roosevelt's re-election campaign and Hitler's rise to power in Germany. If the subjects were unable to predict at a better rate than chance then it would be expected that they would be right in only 50 per cent of cases. In fact a majority of subjects predicted the right outcome for all the selected events. A similar study was carried out by Cantril (1938) in which a sample of 205 middle-class US subjects were similarly asked to predict the outcomes of a series of contemporary events. Loye looked at Cantril's findings in 1976 and checked to see if the predictions of this group held out over time. In his 1995 paper Loye reported that the Cantril group successfully predicted events spanning the highly changeable times around the 1914–18 and 1939–45 wars 64 per cent of the time. Loye extended this line of work with his own study in 1983 when he surveyed 1,500 people drawn from thirty-three US states and representatively sampled for age (16–70), gender and employment. He then selected a sub-set of subjects who showed equal levels of right and left hemisphere cognition (a factor he hypothesised would be important in determining outcomes). From an 11-question prediction test which had one follow-up question in the subsequent year, he found a prediction success rate of 66 per cent.

How could this type of prediction take place? Well, Loye has a series of propositions which are of considerable interest to any general debate, but the most relevant to this current analysis is what he terms 'ideological matrixing' (Loye 1995: 25). By this Loye means that from being immersed cognitively in the prevalent ideologies of the day, some people were able to synchronise with ongoing patterns of change and thus predict outcomes which were formally uncertain. This concept may be clearer if you recall the idea of 'zoning' introduced in chapter four, where, like Lewis's bear, we function seamlessly with ongoing events in the external world. We do not calculate our participation in a game of football or a boxing match, our involvement during these moments is instead transcendent. That as a species we have the ability to 'zone' is difficult to refute, it is only extending this concept to include social as well as behavioural events which is new and contestable.

These may seem to be problematic claims, but if we recall the arguments about the development of adaptive rationality, then the possibility of extending this to the recognition of patterns in social events should not seem so unlikely. Recall that even within the physical sciences it is recognised that there exists order within chaos, that is, that it possible to average close trajectories in a chaotic system to arrive at a state of knowledge of the system from which predictions could be made. All that is being stated here is that it is reasonable to suppose that during the course of evolution the innate ability to discern order in chaotic systems may well have been selected for through natural selection. If so, then this process could occur by way of an algorithm which is related to a formal summing of trajectories. If nature contains inherently chaotic, non-linear and aperiodic events which impact directly on human survival, than it is likely that we will have evolved algorithms which enable us to discern order in such systems. If this is indeed the case, then the evaluation of patterns which arise within chaotic systems will lead to useful knowledge about underlying structures.

VIEWING A CHAOTIC WORLD

What this chapter has revealed is that although non-linear systems may display chaotic behaviour, this does not mean that useful knowledge cannot be gained about underlying structures and possible future events. Clearly our knowledge about such systems is going to be limited when compared with linear systems, but nevertheless, as with the weather, understanding can be gained from mathematical modelling or through running simulations with Boolean networks. By these means it may be possible to gain knowledge about possible future events – and such understanding may not have to be of only a quantitative kind. Through observation of dynamic patterns which can be generated by certain simulation programs of non-linear systems, it may also be possible to uncover new knowledge. Indeed, it may be that human reasoning is particularly well adapted to discern patterns amidst chaotic events. It seems beyond question that although increasingly it is demonstrated that psychological, social and cultural events may have complex or chaotic structures, we can nevertheless make sense out of what is happening. Although we may not be able to predict events with great accuracy, seldom do events within our own culture come as a great surprise. The processes of discerning patterns in a chaotic world may be imprecise, but it would seem that they will allow broad trends to be comprehended.

What we now need to do is determine what these techniques and insights can tell us about why some people drink more than others, and why some encounter a range of personal and social problems through their use.

Chapter 9

Chaos and intoxication

One of the main problems facing anyone who seeks to understand alcohol and illicit drug misuse is the plethora of different ideas which comprise this field of study. There are many different approaches, from disease models which claim that substance problems arise from biological or psychological dysfunction (see Jellinek 1960), to ideas about the role of social learning, where the user is deemed to have incorporated use into a general lifestyle through patterns of association, to the depiction of such problems as being akin to an uncontrollable desire, like a form of obsessive eating (see Orford 1985). Each of these different perspectives, and the many, many more which have not been cited, offer valuable insights into the condition and experience of problematic substance use, but none, unfortunately, accounts for the wide range of data there are on the exact particulars of use. That is, although there have been many attempts over the years to produce a theory of substance problems, or addiction, no one theory can account for the range and diversity which exists with respect to the nature of such problems. Any summary look at the field will reveal that many different components of substance use have been identified. Although some researchers may favour, for example, psychological, biological or cultural explanations, the empirical reality is that each of these areas has made important contributions to the field. Research findings have illustrated that biological, genetic and psychological factors play a part in the emergence of problematic substance use. In addition, behavioural components such as anti-social and delinquent activities, demographic factors such as gender or ethnic origin, and environmental determinants arising from within the family or peer group have all been found, at one time or another, to contribute to the incidence of substance misuse. But despite this abundance of empirical research we are little further on in the pursuit of a general theory of substance use than we were when Jellinek published his influential text in the 1960s. In general terms, although we now know a great deal about the biological, psychological and social elements which comprise substance misuse, there has yet to be developed a theoretical perspective which can account for this diversity of empirical data. In essence, the current problem in drug and alcohol studies is not a lack of empirical data, but rather an absence of theoretical development.

This is not surprising as the task of incorporating all the known facts and perspectives on alcohol and drug problems into a general theory is profoundly difficult. Indeed, it may well be the case that human activities are so diversely emergent that we may never see a time when a complete understanding of any aspect of ourselves will be achievable. Nevertheless, there is no need to be pessimistic. Although we may never gain a complete understanding of ourselves, this does not mean that new syntheses of existing empirical data cannot afford new forms of understanding. Recall, as discussed in chapter eight, that even an appreciation of the patterns of events in human lives can afford useful knowledge to a species within which a pattern-oriented adaptive rationality has emerged. Until recently even the achievement of this modest goal, a view of the deep pattern of substance misuse, will have been beyond current knowledge. But of late the new science of chaos theory has provided a means by which seemingly unknowable patterns of change, like those featured in weather systems and other non-linear dynamical systems, can be rendered meaningful. Sometimes this new perspective can offer no more than a glimpse at a fractal pattern and thus provide an intuition into the underlying structure of complex natural events. However, at other times, through the use of emerging mathematical approaches such as the Lyapunov exponent, knowledge gained can be systematic and repeatable so that others can verify the results.

This book has been an attempt to sketch out what chaos theory can offer to drug and alcohol studies. At times it may have seemed that everything but substance use was being considered, but this was necessary as the first task was to outline in sufficient detail the range of current findings which needed to be considered. That is, the task to be faced at the beginning was to present a discussion of the findings that the reader needed to be familiar with to be able to grasp the contribution that chaos theory could make to this area. However, this needed to be more than merely presenting some of the latest findings and then interpreting these in relation to chaos theory because chaos theory itself is not the answer to the question of why some people use more than others. Chaos theory provides a way of understanding the complexity and unpredictability of use, but there must be another level of meaning. If human action is intentional, then it needs to be intentional towards some object. Chaos tells us how certain patterns form, but not why. The reasons lie elsewhere, within the realms of natural selection. As Stuart Kauffman has pointed out, order emerges from chaos through the processes of natural selection. Hence, any theory of problematic drug and alcohol use needs to explain adequately the presence of ordered intoxication within our society, and how, for some, this becomes disordered or problematic. Consequently we needed to take a long journey through some difficult conceptual territory before we arrived at a point at which the general question could be addressed. So we have arrived, by way of the neurophysiology and evolution of the brain, behavioural genetics, natural selection, adaptive rationality and non-linear dynamics, at the beginning of the final task: the construction of a chaos theory of intoxication which incorporates the concept of natural selection.

Two of the central aspects of a chaotic structure are the presence of sensitivity to initial conditions and scaling. Scaling is a fundamental property of fractals that refers to the way a primary structure repeats over a range of scales. Select one part of the structure and expand it and a duplicate pattern of the whole structure will be revealed, as with a coastline. So, if chaos theory is to be found to be relevant to a theory of intoxication, then scaling needs to be present: sensitivity to initial conditions needs to exist at each level of the phenomenon. That is, it will need to be demonstrated that patterns of intoxication as a whole are sensitive to initial conditions, and also that each of the component elements at the biological, environmental, cognitive and social and cultural levels are similarly dependent upon initial configurations. The discussion so far suggests that this is indeed the case, as the following discussion of these components of intoxication will demonstrate.

NEUROPHYSIOLOGICAL RESPONSES TO INTOXICANTS AND SENSITIVITY TO INITIAL CONDITIONS

There can be little doubt that some of the earliest events in the processes which lead to the production of a human being are crucial in determining the form of the emerging person. The sexual and developmental processes which govern the morphogenesis of each individual, with the exception of identical, monozygote twins, will differ from each other in many varied respects. Through pioneering research by Mendel and continuing developments in contemporary genetics we now know that the recombination of genes during sexual reproduction can give rise to vast diversity in form within the human species. Even the smallest changes in any one gene on the 46 chromosomes which each of us have can produce a large change in outcome for the individual. The existence of severely debilitating disorders which arise from the pairing of certain single recessive alleles is clear testimony to this.

Usually, the differences which arise are not as large as the difference between generally good health and severe illness. However, these findings show that small dissimilarities in genotype can give rise to very small differences in metabolic pathways, which may in turn lead to important disparities in outcome with respect to large-scale phenomena, and result in significant differences at the behavioural level. Recall that there are approximately 12 billion nerve cells in the average human brain from which arise more cellular interconnections than there are particles in the whole of the universe. Clearly this is a highly complex system within which it is reasonable to assume that any small change in the genetic template from which the morphogenesis of the brain commences, prior to the selective events depicted by Edelman, could lead to important differences in neurophysiological structures.

Chapter one provided a detailed discussion of the neurophysiological processes which underlie the metabolism and experience of intoxicating substances. These processes were seen to reside mostly within the brain stem and limbic regions of the brain and to involve particular neurotransmitters such as serotonin, dopamine and gamma aminobutyric acid. Taken together these transmitters have been found to be closely associated with mood, sleep, pain, emotive arousal, motor control and sensory processing. It is not surprising, therefore, that changes in the levels or action of these transmitters have been found to affect the experience of alcohol and other mood-altering drugs.

Traditionally, the focus of research into the experience of alcohol and drug use has been the importance of negative symptoms of withdrawal in sustaining use, but more recent studies draw attention to the equal importance of craving and euphoria. That is, a great deal of research, only some of which has been reviewed in chapter one, has drawn attention to compulsions to use and also the pleasure which use can impart. Drug and alcohol use is thus not merely about negative reinforcement, but also involves positive reinforcement. Additionally, not one but many different sites of action may be involved in the processes that underpin the experience of alcohol and other drugs. These findings reveal that the mechanisms that give rise to positive or negative experiences of alcohol or other drug use may be quite distinct, and that in consequence the experience of pleasure from use may arise from a quite different basis than does continued use to avoid withdrawal.

Whether positive or negative reinforcement is most important in sustaining use is, of course, difficult to discern. On balance it is probably the case that with respect to each person there is a unique resolution of these differing components of use such that each person has a different neurobiological basis from which motivation or direction stems. In other words, the use of alcohol and other drugs arises differently for each person from genetic differences which determine the particular effects of given substances upon particular neurobiological sites. In turn, diverse patterns of regulation of the action of neurotransmitters may give rise to a balance of positive to negative reinforcement which is particular with respect to given substances, and varies between individuals. That is, positive and negative reinforcement will exist in different combinations for different people with respect to different substances. As discussed in chapter one, the way a substance is experienced will arise from the resolution at the level of neuro-physiological processes of the specific neural responses which mediate the formation of a complex reinforcing event. This event will be unique to each individual. The individual differences arise, of course, at least in part, from the recombination of chromosomes from each parent. This recombination is the main but not sole basis for such phenotypic differences.

Of course differences which arise in phenotype are not only determined by recombination during sexual reproduction. Gene mutations can also take place in germ cells and will thus be passed on to offspring, leading to phenotypic changes. And changes need not only arise from the effects of single genes. Work with QTLs (Quantitative Trait Loci) has shown that multiple genes can interact to produce certain quantitative outcomes. In fact certain quantitative traits can stem from the interaction of different QTLs.

Gene expression also plays a part in adding to the complexity of these events. As previously discussed, gene expression can be controlled by environmental events such as the levels of a specific protein. Varying levels can act to switch on and off the action of specific genes, which may in turn affect the action of other genes, which in turn affect changes elsewhere, and so on in a complex multi-dimensional pattern of interdependence. And as Ruelle (1993) observed, the more dimensions there are in a system above two, then the greater likelihood there is that chaos will be encountered.

Through describing these processes it can be seen that the events through which an individual acquires a predisposition towards intoxicants is not a simple linear process which leads necessarily to a set of particular consequences. The evidence available suggests that at each stage in the formation of a neurophysiological response to alcohol and other intoxicants there is considerable complexity and uncertainty. Differences stem from multiple events which include the effects of genetic recombination and mutation which subsequently result in the modification of gene expression through QTLs. Any initial differences which arise will be increased to a marked degree during successive iterations of developmental processes, through feedback from the environment at both micro and macro levels of development. As will be appreciated from the discussion of chaos theory in chapter seven, the only way to be able to predict what will develop from a non-linear dynamical system such as that depicted here is to know exactly the initial configuration of the system. If this were possible, and if we also understood the exact mechanisms through which the consequent developmental process operated, we might then have the means to predict outcomes for the individual. However, we do not have such knowledge nor are we ever likely to, as changes even at the quantum level will eventually appear at the micro level, as discussed in chapter seven. In consequence, there arises for each individual a distinct profile with respect to the neurophysiological structures which underlie the positive and negative reinforcing events, which cannot be known in advance of its emergence. And it would seem to be the case that in consequence of these characteristics, responses to intoxicants at the neurophysiological level are indeed sensitive to initial conditions.

GENETICS, THE ENVIRONMENT AND UNCERTAINTY

The case for a biological basis of alcohol and drug problems has been made clearly by the research evidence presented throughout the early chapters of this book. It is now widely appreciated that substance problems are not only a social and psychological phenomenon; each of the primary components of misuse which include withdrawal, tolerance, craving and euphoria have been found to be linked to specific sites of action within the brain. The interaction between substances consumed and neurophysiological processes is not simple, however, as the complex nature of the biochemistry of the brain, structured by way of non-linear developmental schemata, means that responses to intoxicants will differ between individuals. Nevertheless, despite this variance between individuals, such differences are not randomly spread through a population, but are instead distributed according to the laws of inheritance. These have already been referred to as one source of uncertainty of outcome with respect to the neurophysiology of the brain. There are yet further aspects of inheritance, however, which require review with respect to the occurrence of diversity and uncertainty within biological form.

In chapter one reference was made to the observations of Crabbe and Goldman (1992) concerning the finding that children who grow up in households where there are alcohol problems have increased chances of either developing problems themselves, or becoming abstinent. This was cited as being an important finding in that it drew attention to the fact that a person's genotype does not exclusively determine that prevalence of complex traits such as alcohol or substance problems. There thus seemed to be a strong case for arguing that other developmental elements were important in determining outcomes, and that amongst such additional factors would be environmental influences. To review this possibility attention was given to the work of Robert Plomin and others on the relationship between genetics and the environment.

What current research into this relationship between nature and nurture has revealed is that many environmental measures such as marital, financial, health and employment difficulties (Holmes 1979) have been found to have genetic components. In addition, work has been carried out on the genetic basis of psychological traits, and here also degrees of inheritance have been found. Plomin (1994) has argued that it may be the case that the covariance found between environmental measures and traits such as cognitive abilities or psychopathological dysfunction may have genetic components. Indeed, genetic effects have been found with respect to the relationship between life events and personality and socioeconomic status and intelligence. Hence it seems that genotype and certain environmental measures may be mediated by psychological traits in that genes identified as being associated with measures of the environment will in part also be identified with psychological traits (Plomin 1994).

However, although genetic components have been found through calculating heritability statistics (a measure of the covariance between a phenotypic and environmental measure) for a range of traits and environmental measures, one important finding is that no genetic component has been found to sum to 100 per cent. That is, even where genetic components of environmental measures have been found, experience (nurture) has also been found to be influential. This may be an unsurprising conclusion, but the fact that measures such as the selection of a peer group arise from both biological and social or cultural influences is of considerable importance. Nevertheless, these studies point to the importance of nurture as much as nature in determining outcomes for the individual. And in this respect, studies on the importance of the non-shared environment encountered during early childhood may be important in unravelling the complexity of individual difference. That is, attention needs to be focused on differences in the way siblings, including twins, are treated in the home. As Reiss *et al.* found, such differences can be highly influential in effecting dissimilarities between siblings in later life:

> a sequence of nonshared experiences . . . may protect some siblings in a family, but not others, from becoming alcoholic when one or both parents are alcoholics. Early in life the sibling who will grow up free of alcoholism is protected from alcohol by the family group; the family keeps the family practices most

cherished by the child free from intrusion by the alcoholic behavior.

(Reiss et al. 1991: 288)

So it seems that genetics is important in determining both differences and similarities. We inherit our genotype from our parents and this plays a large part in determining phenotypic outcomes. However, it is not the case that from the moment of fertilisation onwards developmental processes proceed by means of an exact template, even between identical twins in some respects. Recombination and the results of the stochastic processes of protein synthesis, governed by complex feedback mechanisms between genes and the larger biological environment, inevitably lead to uncertainty of outcomes, as described above. Now, added to this is the further effect of the social and cultural environment on the phenotype. As studies on the relationship between nature and nurture have shown, only part of the variance in psychological traits and environmental measures can be explained by genetic influences. Also important are the differences which each person encounters in the external environment. What we do not share can be as important as what we do share in determining what we may become. Even small differences in the treatment siblings receive from the adults responsible for their upbringing can give rise to considerable differences later in life. Current research evidence suggests that differences in treatment may be quite marked for some measures, such as negative behaviour, even between same-sex siblings (see Reiss et al. 1991).

Consequently, there appears to be increased complexity as biological and social development takes place. At each stage of development the direction that is taken will be dependent upon past events as developmental processes show sensitive dependence upon initial conditions. Edelman's perspective on the morphogenesis of the brain, within which he draws attention to the processes of natural selection which are involved in determining which cell lines will propagate, is central to this idea. However, they will also be subject to restructuring by continuing experience. Unlike a weather system, which is fully formed in terms of all its components, biological life is developmentally governed by a sequence of events which constantly change the composition of the whole system. The simple biological system which is a three-day-old fertilised egg is not the same biological system which exists with respect to a 32-week-old fetus, even though one has formed from the other. Consequently there exists with respect to biological development a meta-complexity whereby complex systems which display sensitive dependence upon initial conditions are evolved into other complex systems which also display sensitivity through the action of an external environment. Given this nested complexity, it is not surprising that there can only be great uncertainty with respect to the prediction of outcomes. In effect, we really should not be able to predict anything about an individual; whether, for example, they will become a great composer, or a homeless person dependent upon alcohol. Human life does indeed have the characteristics of a chaotic system. However, this is only part of the story. Although we are not able to tell at the level of the individual whether or not they will become homeless or rich and famous, we can make broad predictions as to the kinds of people they will become. In other words, there are species-level constraints on what is possible, and there are also cultural-level constraints on how species-level elements may be expressed or realised. And let me be clear about what is being said here. Given the stochastic events which underpin biological development and give rise to a meta-complex system which is highly dependent upon initial conditions, human existence is potentially chaotic. However, observation leads us to conclude that this is not the case routinely. Hence there must be other processes which underpin development such that recognisable biological and social order arises. How this occurs, how order emanates from chaos, is of course by way of natural selection. Hence it is being claimed here that it is the selective action of the environment, acting through experience, which gives rise to the ordered self. However, this does not mean that an ordered state will necessarily endure with respect to all aspects of this ordered self.

ORDER FROM CHAOS

The basic conditions of sexual reproduction and protein synthesis as detailed above, if interpreted from the perspective of chaos theory, describe a non-linear system which has sensitive dependence on initial conditions. As you will appreciate, this means that human morphogenesis is at least partly structured chaotically. However, we know from even the most trivial observations that in many respects order rather than chaos seems to best describe most aspects of human existence. We possess an ordered biological form and, in general, human affairs, both psychologically and socially, are ordered into certain recognisable patterns. How order arises within the context of complexity has been discussed previously in chapter 9. Stuart Kauffman's work with Boolean networks has shown how changes in the parameters of such networks can facilitate changes from chaos through order to systems with many unconnected attractors. From studying such phenomena, Kauffman has argued convincingly that at least some natural systems exist at the border of chaos and order. That is, islands of order within a system are surrounded by unfrozen regions, with a boundary region between which is transitional. Thus there are described three possible conditions: ordered, unfrozen or disordered, and the boundary in between (Kauffman 1993). Although research is still continuing, Kauffman places his confidence in the hypothesis that complex adaptive systems reside within the boundary region, at the so-called edge of chaos.

This seems a reasonable hypothesis if an assumption is made that natural systems will need to attain a certain level of flexibility. A system is more likely to be able to respond to external influences if it retains a flexible structure. But there would need to be limitations to this as too flexible a system would not impart the constancy of pattern which coordinated and intentional action would require. Hence, if you were designing an adaptive machine you would want certain parts to be flexible and adaptive, but others, probably associated with essential tasks, you would wish to be relatively fixed. Which were fixed and which remained flexible would be decided by way of natural selection. In Ashby's procedure for designing a brain, the processes he described would fit with this scenario: malleability is maintained until an appropriate configuration of the system is achieved.

There are strong parallels here with the views of Edelman and his perspective on the selection of neural structures by external events. Within his model, through the initial existence of a variety of different neurological structures within the brain, which have arisen collectively as an adaptive trait, the child has a wide range of possible neurological, and thus behavioural, responses, but it is only those which impart an advantage which are selected. You will recall the example of seeking and consuming water which was given in chapter three. This suggests that at particular stages in the development of the child towards becoming a consciously aware individual, there are parts of the physical brain where the structure of the mind is effectively on the edge between chaos and order. In principle these areas, from which adaptive responses may occur, could be wholly chaotic. However, Kauffman's work suggests that the boundary region between chaos and order has the greatest potential for adaption. [2,]

This distinction between the brain and the mind is significant as it conveys the concept that, although it is the complex morphological structure of the brain which provides a physical medium for adaptive cognition, it is within the diverse neural and global maps that order or chaos resides. Although neural and global maps have a physical structure, it is the cognitive, mental outcome which is either ordered or chaotic. Brains are complex adaptive organs of very large dimensions, but it is minds which display order or chaos. To put it another way, the Boolean network which resides on a complex scientist's computer may have chaotic and/or ordered regions. These are descriptions of the Boolean network, not the hardware within which the network is installed. Hence, although they approach the task of describing natural selection and adaptation in natural systems differently, both Kauffman and Edelman are in agreement with respect to the view that order within very large-scale parallel-processing systems arises by way of natural selection.

There are many ways of interpreting these findings with respect to human existence, and some of the latest results from the study of natural selection and human cognition have been presented above, particularly with respect to the significant work of Leda Cosmides. But the intention here is not to rehearse these findings, but rather to say something about what they may contribute to understanding alcohol and drug misuse. Well, the links are conceptually straightforward, though the empirical task of verifying them is more problematic and may thus take some time to achieve.

As has been discussed, research on the relationship between genetics and the environment has shown that many psychological and behavioural traits have genetic components. This is important, as you will no doubt appreciate, because it clearly indicates that patterns of drug and alcohol use may also be partly inherited. However, the reverse side of this is that such traits are not wholly inherited, but also arise from social or cultural experience. No longer can either natural or social scientist claim total jurisdiction over this area. Now, although this finding produces a much-needed realignment of interest in the subject, it also raises problems.

If experiential factors facilitate alcohol and drug misuse then how does this process take place? If we are to avoid the pitfall of reproducing an endless list of variables such as, for example, the home environment or peer groups, or indulging in theorising with unproductive ideas such as social learning, which never adequately explains why a harmful event somehow becomes a beneficial one, then a theory needs to be developed which is consistent with other findings on human action, from biological to cultural perspectives. The key to this lies with natural selection.

NATURAL SELECTION AND DRUG AND ALCOHOL MISUSE

The idea that natural selection plays a part in the formation of drug and alcohol misuse stems from three main propositions. First, the structure of the brain is such that a complex mind evolves which displays both ordered and chaotic regions. Second, the emergence of ordered behavioural and psychological traits is a precondition of coherent engagement with the external world. Third, it has been demonstrated that natural selection is the means through which natural systems are meaningfully oriented towards the external world. Together these add up to an interesting hypothesis perhaps, but are there any empirical findings which support it? I would suggest that reflections on the unshared environment may produce some interesting ideas in support of this hypothesis.

The paper by Reiss *et al.* (1991) drew attention to data which shows that there may be important differences in the early experiences of even same-sex siblings. Now, if this is so, if it is the case as Reiss *et al.* suggest that the individual micro-climate of experience we all have as children is unique, then it means that the selective processes which Edelman describes as taking place from the moment of birth, and which select the neural pathways which underpin our cognition, could result in one child valuing an object or experience, say, alcohol, more than another. The range of events which could elicit such a response could be infinite. They could be as simple as learning to sip a glass a beer to placate an angry parent and subsequently valuing alcohol as a means of achieving a sense of security, or they could be more complex than this and involve more subtle selective forces, such as acquiring status through association with adult social forms. The assumption being made in both of these cases is that both 'security' and 'status' are forms of *primary selective categories.* These are adaptive elements in that the degree to which they are achieved will determine the likelihood of survival or significantly improve the well-being of the individual.

If natural selection can be seen as a basis for the emergence of such traits in early childhood, then can this also be the mechanism through which alcohol problems arise later in life? The answer is probably yes. Recall once again Ashby's concept of designing a brain. The system he designed achieved a configuration which either did or did not achieve its function. If it did then fine, the system stayed as it was, but if it did not meet its function then parameters of the system were changed so that a new configuration arose. This process occurred repeatedly until a satisfactory configuration arose. Kauffman equated this concept to the changing of parameters within experimental Boolean networks. Changing the number of connections between elements or the functions which determined the nature of the connections at random would give rise to changing configurations of the network. Sometimes there could be chaos whilst at other times order arose. Now, it is far from fanciful to suppose that if this is a desired characteristic of very large parallel-processing networks which are subject to natural selection, then it is likely to be a feature of the human mind. The consequences of this are that if the parameters of our cognition change through some events which have destabilised the processes of cognition then the system will reconfigure. It may reorder or it may become chaotic. Such events could be, for example, physical, such as brain damage after an accident, or they could result from emotional shock. The number of different causes may be quite large. Nevertheless, whatever the cause a reconfiguration may occur in many cases. Such reordering could introduce new behaviours such as alcohol misuse, or it could result in a chaotic state which may be displayed as a mental dysfunction. It is impossible to resist making the observation that drug or alcohol use which has become highly problematic and multi-dimensional (involving housing, relationships, financial, criminal justice and perhaps many other components) is described as chaotic drug or alcohol use by many clinicians in the field.

Essentially, therefore, what is being claimed here is that through the action of natural selection at each level in the genesis and continuing life of an individual, adaptive traits arise which fit the circumstances of a person's existence. Some such traits may ultimately be harmful within the broader context of the person's life, for example they may cause ill health, but nevertheless it is proposed that very many behavioural and psychological traits arise because they fit the context of some immediate experience. As unlikely as this may seem at first encounter, it makes sense in terms of species-level survival. In certain circumstances, say due to an environmental disaster, it may be necessary for individual members of a species to engage in acts which may cause them considerable harm or place them in danger. Drinking highly polluted water or foraging very close to a predator may be two examples of such an outcome. However, the consequence of not doing so – dying of thirst or hunger – would be considerably worse than drinking or foraging and taking the risk. Although many members of the species may die as a result of doing so, some may have immunity from the contaminants or be more able to avoid the predators than others and thus survive to reproduce. So it is an advantage for an individual if the means exist within a species for traits to arise which may lead to harmful actions at the individual level. Natural selection thus does not only act through the selection of purely positive traits for the individual, the selection of negative traits is also possible. Indeed, remembering that due to reentrant mapping within the structure of the brain what is selected for is valued, then negative traits which are selected will also have value. The actions proceeding from such traits will be positively valued by the possessor of the trait. And given that humankind is a collective species, such experiences of positive value may of course become celebrated socially, and thus form a collective basis for continued action, which may also draw in other participants. This possibility leads to a consideration of the social components within this analysis, which will be featured in the following section.

UNCERTAINTY AND SOCIAL SPACE

In a recent paper on the social and spatial aspects of illicit drug use (Dean 1995b) attention was drawn to the complex antecedents which appeared to form the foundations of use in rural communities. Fieldwork conducted in rural East Yorkshire and the Western Isles of Scotland illustrated that outcomes related to drug use by young people were not simply a matter of personal choice, peer groups, personality or some other such variable commonly used to describe or depict antecedents to drug misuse, but instead arose from a complex resolution of a number of different factors. The data, which revealed a variety of novel forms of drug use, including the use of pig tranquillisers, pointed to the existence of a multi-dimensional system of effects within which forms of intoxication took place. The primary dimensions of this system were geographic remoteness, social proximity, the influence of new social ideas, individual agency, the history of intoxication within the community and material preconditions.

Geographic remoteness referred to the distance between the young person's place of residence and sources of supply. The more remote the location then the less likely it was that they would be able to acquire drugs or sustain illicit drug use. Social proximity depicted the social distance between an individual and a source of supply. A person might reside in a remote community in northern Scotland, but if they had contacts in London who would be willing to post supplies to them, then they would be able to sustain a regular if not frequent pattern of use. The influence of new social ideas would include vicarious experience encountered via the media, but referred more specifically to the effects of incomers into a community. A new person arriving in a generally self-contained community can have a large impact on social practices. Individual agency represents personal predispositions towards use, a topic already subjected to critical discussion. The history of intoxication within the community is also of importance in shaping patterns of intoxication. The differences in traditional patterns of intoxication – for example coca use amongst native South American peoples compared with the use of alcohol in Hebridean culture – provide an example of the representation of tradition in contemporary forms of intoxication (see Dean 1995a, in press). Of course, the continuing globalisation of many cultural forms may reduce the contribution of local histories to patterns of intoxication. Material preconditions merely refer to the ability to resource consumption.

Given that it can be expected that each of these elements will exist to a unique extent for each individual, and that changes in one element, such as moving to a less remote location, may lead to substantial changes overall, it may be said that such a system is sensitive to initial conditions. In fact the analysis argued that outcomes for an individual stemmed from a resolution of these contributing elements at a particular moment in time; a situation which can, of course, be described by use of key concepts from chaos theory. That is, the status of the individual with respect to drug use (in this case) can be considered, in the terminology of chaos theory, as the point within the phase space of the system which describes the state of the system. As time passes the individual will change their pattern of use, even if only to a small degree on some occasions, according to the configuration of the attractor within the system. Thus, from this perspective, drug use is effectively the result of a time evolution in n-dimensional space which has sensitive dependence on initial conditions. What this means is that patterns of drug use are described by motion on a strange attractor; which also means, theoretically, that even at the social level the outcomes of drug and alcohol use are uncertain at each stage. At the level of the individual it cannot be known in advance what their pattern of use will be during the next time interval. All we can know for certain is that it may change or stay the same.

However, there is a caveat to this, as no doubt you have realised. Within natural systems there is a predisposition towards the establishment of order. As we now know, evolution occurs by way of natural selection acting at the border of chaos within complex systems, and that from this, ordered systems emerge. It would be expected, therefore, that if natural selection does act upon social behaviour then this condition would apply to social drug use. Well, from the research of Leda Cosmides we know that natural selection does act to affect certain outcomes within the context of the rules of social exchange. Hence there is good reason to suppose that it may be possible ultimately to demonstrate a wider applicability of natural selection with respect to other aspects of social action. In fact it was hypothesised in chapter five that there may be a deontic logic with respect to social exchange with children. In the case of drug and alcohol use it also seems reasonable to hypothesise that ordered forms of the pursuit of intoxication, a practice which seems universal within human communities, have evolved through natural selection. But, as with the case of systemic changes which may occur for the individual (the reconfiguring of a system through changes in inputs or functions), changes at the social or cultural level may also lead to departures from an ordered state. But a chaotic social state will, through natural selection and the survival imperative this implies, eventually settle down into a new ordered state. These emerging states may not represent a universal rationality of use, but may instead reflect an adaptive rationality which is closely centred on the experiences of the individual. Indeed, in the postmodern world where reason and formal rationality no longer have an institutional resolution, it may indeed be expected that a plethora of adaptive rationalities will form. As such, diversity will be the consequence of natural selection, acting as it does through the medium of competition at the level of the individual, and outcomes favoured by tradition, or of unremitting worth to each individual within the community, collectively or individually, may not necessarily arise.

It would thus seem to be the case that sensitivity to initial conditions exists at each level of alcohol and drug use, from the biochemistry of the brain through to society and culture. At each level outcomes cannot be predicted. However, although this describes the existence of chaos, the general rule is for order to arise through the processes of natural selection. Nevertheless, the existence of chaos at each level of human life suggests that existence can be described by a fractal structure – an adventurous proposition, but can it be true?

Chapter 10

Is existence fractal?

What the preceding analysis has revealed is that what we become cannot easily be determined. Indeed it is usually the case that, notwithstanding the fact of order in our lives, life follows a very uncertain and unpredictable course. This is not because events outside our control are only conditional, and thus subject to change, but because the nature of human existence is complex, adaptive and balanced between order and chaos. The order that we observe daily is a variable condition. In some cases order appears to persist over very long periods, whereas in others change seems to be an almost continuous affair. This is because we are not single-layered, with each feature of our existence occurring on the same plane. Instead, as discussed above, our presence and form is the outcome of a meta-complex framework, each layer, from biochemistry to culture, being represented by a dynamical system which has sensitive dependence upon initial conditions. However, at some levels within this framework order tends to predominate, whereas in others change is a more frequent event.

With respect to the predominance of order, there are species level constraints on what is possible, so change here would have to proceed extremely slowly. Developing the facility to breathe air did not, for example, take place overnight for previously water-based life-forms. In this regard we can expect that there are many features of our existence which will change only slowly, if at all. What these may be will be difficult to predict. And over very long periods of time, measured in millions rather than thousands of years, changes in our form and consciousness that may appear mere fancy now may indeed come to pass. However, such long time-scales need not concern us on a day-to-day basis. For most practical considerations, our macro-biological form, such as the broad structure of our brain or heart, is a fairly fixed property.

This is not the case, however, with regard to other aspects of our being. As Edelman's work has shown, changes in neurophysiological structure can take place during an individual's lifetime in response to experience. In consequence of this malleability, each of us may see the world in a different way. So, the uncertainty here is greater than that which exists with respect to the broad structure of the brain. Not only will we have slightly different networks of neural pathways from each other, but, as a result, we will most likely also see colours and hear sounds differently. Of course this malleability is temporary as, after a finite period of development, these aspects of the neurophysical networked structure of our brain eventually become generally fixed into specific forms and, barring a congenital defect or physical damage, remain that way.

At another level in the hierarchy of the human form order also arises, but malleability remains a more permanent feature. This is the level at which neural networks, or maps, which reentrantly interconnect various structures of the brain to achieve specific biological and behavioural functions, are interconnected to form global maps. In linking the cortical areas of the brain with regions concerned with speech and value, global maps provide the underlying structure for the emergence of conscious, intentional action. It is at this stage, at the level of symbolic consciousness, that Cosmides's Darwinian algorithms can be said to be grounded. That is, Darwinian algorithms are a form of global map (hence, as argued in chapter five, adaptive rationality will have a value component). Given that Darwinian algorithms are the means by which we consciously engage with the social world, then it should be expected that at this level greater malleability will exist. This is not to say that Darwinian algorithms themselves are subject to change. I am fully convinced of Cosmides's reasoning with respect to the origin of the laws of social exchange being from within the context of the social lives of our hunter-gatherer ancestors. In this sense, the Cosmides algorithms of social exchange are an example of enduring order in cognition. What I do claim, though, is that the algorithms which exist at any moment in time do not encompass all cognitive potential at the level of adaptive rationality. Amongst the complexity of neural maps there is the potential for new global maps to arise from natural selection, and indeed, it is proposed, new algorithms do form on a continuing basis. It is, in fact, hard to imagine the mechanism whereby the malleability which existed prior to the emergence of the laws of social exchange, and which was a condition of such an emergence, could have subsequently declined. Of course, as far as we can tell, evolution can reach an end point; for example, the cockroach which has been unchanged for some 1 million or so years. However, there is nothing to suggest that this is the case with respect to human cognition. Recall that even Cosmides's work did not show that all participants completed the problems of social exchange in the way the theory predicted. Such variance suggests the possibility of the existence of other algorithms. Indeed, looking for exceptions to linear rules is one way of seeking out an underlying chaotic structure.

The view that adaptive reasoning is ordered but not fixed, and is subject to natural selection in the contemporary world, is one of the central ideas underpinning the argument for a chaos theory of intoxication. This is so because by these means it can be understood how practices seemingly undesirable to the individual come into being, and how, having done so, they are inherently unpredictable in nature. Clearly all but identical twins are different at the biological level and this is the first level of sensitivity to initial conditions, but even greater dissimilarities stem from the infinite differences which exist within the realm of experience. Evidence from studies such as those of Plomin (1994) and Reiss *et al.* (1991) illustrate that the non-shared environment is at least as important as any other developmental factor in influencing outcomes for the individual. Any small event happening at an early stage in the formation of the conscious self, and through later stages of socialisation, could give rise to large differences in both behaviour and adaptive rationality between closely raised siblings. Through processes of natural selection, ways of symbolising the external world and acting within it will arise to fit the nature of experience at the level of the individual. These ways of acting and rationalising will, of course, be subject to species-level constraints, but they will not follow a predetermined blueprint. There will be ways of acting and rationalising which are historical, having been passed on by biological or cultural means (the inheritance of *memes*, which have been suggested as being the cultural equivalent of genes, see Dawkins 1976), but there will also be individual outcomes which will to some degree be unique. It is in this sense that the development of our adaptive rationality can be said to be sensitively dependent upon initial conditions.

I first gained an insight into the importance of this in affecting drug and alcohol use when interviewing a person who, in the past, had had a large and expensive cocaine habit. During this study I conducted life-history interviews in the belief that by doing so I could gain insights into those moments in a person's life which may be thought to be formative in terms of such actions. Of course, I now know that it is not possible to generalise directly from such findings as each person's life is unique; the only pattern being that something happens which leads to subsequent substance misuse. Naturally, there may be more than one thing involved; all we can be fairly sure about is that due to sensitivity to initial conditions, we can never empirically determine exactly the configurations of effects which led to the perceived outcome. Nevertheless, it is a useful exercise in terms of understanding the individual person and being able to describe the distinct patterns of their life compared to a person with similar or related substance problems.

But I digress. I was describing a particular case as an example of previously conducted life-history interviews. This particular person had started out with alcohol misuse, eventually progressed to injecting heroin and moved on to smokable cocaine. He was single, in his late thirties and wealthy enough to manage a £200 per day cocaine habit without having to resort to drug-related crime. At first I was interested in why he had switched from injecting heroin to smoking freebase cocaine, which, in the early 1990s did not appear to be a usual progression, particularly amongst wealthy middle-class users. After many hours of discussion he revealed that he had stopped injecting heroin after a possibly life-threatening experience. He had been with a number of people, one of whom was a close friend, and he and the close friend were sharing the heroin they had just bought together. The friend injected first and a short time later my subject prepared to do the same when he was told that his friend seemed to be in some trouble. Shortly afterwards the friend died from what was subsequently established to have been an overdose. My subject and his friend had bought the heroin together from a source they had not used before, and it appears that the cut had been more pure than they were used to, hence the death from overdose. My subject was very frightened by the experience and gave up both heroin and injecting for what he perceived as the safer option of smokable cocaine. This is not an unusual story as no doubt many readers will agree. Many people with substance problems will not cease use until some event, which on many occasions involves an encounter with the possible fatal outcomes of use, leads them to fundamentally question their drug or alcohol use.

However, this present case history has not yet been concluded. The above events gave an important insight into how sensitive use can be to changing events. A change in the way heroin was perceived led to a switch in usage. The subject's pattern of use was reconfigured towards a different, though similar object and the establishment of a new and ordered pattern of drug use; £200 per day, no more and no less, on crack. But what triggered the original heroin use, and the alcohol use before the heroin use? These were my pre-chaos days and I wanted *the* answer!

We continued with our discussion at different times and in different places until he recited an early recollection of his first encounter with alcohol during his pre-teenage years. His parents owned a drinks cabinet which contained various decanters of alcoholic drinks. He did not know what they were, presumably whiskey, sherry and port, but he did recall the special status they were given within the home. In his reminiscence he recalled that they were invested with an almost magical status; they were not to be touched except by his parents, who seemed to value the contents highly. They were also, to him at that time, beautiful to look at. He described the deep colours and the shining glass formed into attractive shapes. Thus, for him they assumed high status shrouded in mystery and became extremely desirable. So alcohol acquired an exotic and desirable image; it became something which once attained would be a passport into the mysterious and pleasurable world of his parents. A set of circumstances which in themselves are not unusual, led in this case to a preoccupation with intoxication which was informed by a pursuit of the forbidden.

If this sounds rather Freudian then I make no apologies as Freudian analysis also looks for those moments in the processes of acquiring a consciously gendered position in society which give rise to normatively undesirable outcomes for the individual. At that time in the early 1990s I followed a similar path, only to discover endless diversity rather than patterns from which generalisations can be made. Hence the insight into the fact that small events lead to unpredictable outcomes, and the recognition that natural selection and chaos theory provided a means to describe the underlying structure of such infinitely variable events. Which brings us naturally to the level of social and cultural interaction, where at least an equal amount of uncertainty can be confidently said to arise.

As the work undertaken on rural drug and alcohol use in the Western Isles of Scotland and East Yorkshire illustrates, many chance events can give rise to unpredictable outcomes. New friends or acquaintances, moving to a new location nearer or further away from a village, town or city, moving into a new area or country with a different historical pattern of intoxication, all these elements can influence outcomes for the individual. This describes a multi-dimensional, dynamic system within which the individual can be described from one moment in time to the next as a resolution in time and space of all the competing influences; in effect, they are the point in bio-social phase space which describes their status (i.e. the state of the system).

Many of the young people I got to know during the work in the Western Isles eventually moved, at one time or another, to Edinburgh, Glasgow or Aberdeen for work or to attend college. Each of the main participants in the study was a part of the same extended social circle and would have been involved in heavy drinking and/or soft drug or solvent use (see Dean 1990) whilst resident in Stornoway. However, despite these similarities, the changes which took place once they had moved to a different community were dissimilar and unpredictable. One of the academically least well-qualified participants, a person who had been drinking in excess and had been using both cannabis and solvents, took a number of low-paid jobs, eventually gained a higher education qualification, was promoted, got married and subsequently took part only in social drinking. Another member of the group who was relatively well-qualified through matriculation and who never used anything but alcohol, developed a very heavy drinking habit, failed his college course and returned to Stornoway. Some of the young women involved with the study became pregnant and never left the islands, whereas others gained good degrees from Glasgow or Edinburgh and secured careers in the media or health-related professions. From their actions during their time in Stornoway, it was not possible to predict how participants' lives would turn out once they left and settled in a mainland city. Some of the most able or committed failed to succeed, whilst those who seemed at the time to be drifting into unqualified work and continuing alcohol, soft drug or solvent use turned out to be the most successful. As described in detail in Dean (1995b), what happens at the social and cultural level can be best understood as being a resolution of a variety of influences at the geographic, personal, social, cultural and historical levels.

What has been revealed from this current analysis is that there is good evidence to support the argument that from deep structure at the biological level through to the rapidly changing context of social and cultural forms there exists sensitive dependence upon initial conditions. In consequence, despite the fact that we know a great deal about intoxication, it is impossible to predict accurately the course of an individual life in this regard. This is not to say that order does not exist – clearly the evidence is that through the process of natural selection order will arise at each level of existence, from the biological through to the cultural levels. However, such order can and is transcended in response to events which may take place at any level of existence. When this happens the process of natural selection will be the basis or structure from which a new configuration will arise. Always the template will be 'Does this work for the individual in these specific circumstances?' If it does, in the sense that it imparts a survival advantage, then the system will stay as it is at that point. What works is infinitely variable, dependent as it is upon the initial conditions which pertain with respect to the meta-complex form of the individual. We are constrained by the requirements of our species, and order must also dominate overall if life is to continue in an evolutionarily successful way, but at each level of our existence there exists an underlying chaotic structure which is the mirror of each other layer above and below. In this sense human existence in all respects, and not just with regard to intoxication, is fractal. But Darwinian natural selection leads to order, to different extents at different levels of our meta-complex being. In consequence, at the individual level we cannot know exactly where order predominates or where chaos is emergent.

Bibliography

Allgulander, C., Nowak, J. and Rice, J. P. (1991) 'Psychopathology and treatment of 30,344 twins in Sweden II: heritability estimates of psychiatric diagnosis and treatment in 12,884 twin pairs,' *Acta Psychiatrica Scandinavica* 83: 12–15.

Allgulander, C., Nowak, J. and Rice, J. P. (1992) 'Psychopathology and treatment of 30,344 twins in Sweden,' *Acta Psychiatrica Scandinavica* 86: 421–2.

Anthenelli, R. M. and Schuckit, M. A. (1991) 'Genetic studies of alcoholism,' *International Journal of the Addictions* 25: 81–94.

Anthenelli, R. M. and Tabakoff, B. (1995) 'The search for biochemical markers,' *Alcohol Health and Research World* 19.3: 176–81.

Ashby, W. R. (1960) *Design for a Brain*, 2nd edn. New York: Wiley.

Balfour, D. J. K. (1994) 'Neural mechanisms underlying nicotine dependence', *Addiction* 89.11: 1419–23.

Bergeman, C. S., Plomin, R., Pedersen, N. L. and McClearn, G. E. (1991) 'Genetic mediation of the relationship between social support and psychological well-being,' *Psychology and Aging* 6: 640–6.

Bozarth, M. A. (1994) 'Opiate reinforcement processes: re-assembling multiple mechanisms,' *Addiction* 89.11: 1425–34.

Brett, J. F., Brief, A. P., Burke, M. J., George, J. M. and Webster, J. (1990) 'Negative affectivity and the reporting of stressful life events,' *Health Psychology* 9: 57–68.

Brown, T. A. (1996a) 'Measuring chaos using the Lyapunov exponent,' in E. Elliott and L. D. Kiel (eds) *Chaos Theory in the Social Sciences. Foundations and Applications*. Ann Arbor: University of Michigan Press.

Brown, T. A. (1996b) 'Nonlinear politics,' in E. Elliott and L. D. Kiel (eds) *Chaos Theory in the Social Sciences. Foundations and Applications*. Ann Arbor: University of Michigan Press.

Cadoret, R. J. (1994) 'Genetic and environmental contributions to heterogeneity in alcoholism: findings from the Iowa adoption studies,' *Annals of the New York Academy of Science* 708: 59–71.

Çambel, A. B. (1993) *Applied Chaos Theory. A Paradigm for Complexity*. San Diego, CA: Academic Press Inc.

Cantril, H. (1938) 'The prediction of social events,' *Journal of Abnormal and Social Psychology* 33: 364–89.

Cloninger, C. R. (1987) 'Neurogenetic adaptive mechanisms in alcoholism,' *Science* 236: 410–16.

Cloninger, C. R., Bohman, M. and Sigvardson, S. (1985) 'Psychopathology in adopted-out children of alcoholics: the Stockholm adoption study,' in M. Galanter (ed.) *Recent Developments in Alcoholism*. New York: Plenum Press.

Cohen, A. P. (1994) *Self Consciousness. An Alternative Anthropology of Identity*. London: Routledge.

Cook, C. and Gurling, H. (1991) 'Genetic factors in alcoholism,' in T. Palmer (ed.) *The Molecular Pathology of Alcoholism*. New York: Oxford Press.

Cosmides, L. (1989) 'The logic of social exchange: has natural selection shaped how humans reason? Studies with the Wason selection task,' *Cognition* 31: 187–276.

Cotton, N. S. (1979) 'The familial incidence of alcoholism,' *Journal of Studies on Alcohol* 40: 89–116.

Crabbe, J. C. and Goldman, M. D. (1992) 'Alcoholism: a complex genetic disease,' *Alcohol Health and Research World* 16.4: 297–303.

Crabbe, J. C., Belknap, J. K., Mitchell, S. R. and Crawshaw, L. I. (1994) 'Quantitative trait loci mapping of genes that influence the sensitivity and tolerance to ethanol-induced hypothermia in BXD recombinant inbred mice,' *Journal of Pharmacology and Experimental Therapeutics* 269.1: 184–92.

Darwin, C. (1859) *On the Origin of the Species by Means of Natural Selection of the Preservation of Favoured Races in the Struggle for Life*. London: John Murray.

Dawkins, R. (1976) *The Selfish Gene*. Oxford: Oxford University Press.

Dean, A. (1990) 'Culture and community: drink and soft drugs in Hebridean youth culture,' *The Sociological Review* 38.3: 517–63.

Dean, A. (1995a) 'Space and substance misuse in rural communities,' *International Journal of Sociology and Social Policy* 15: 134–55.

Dean, A. (1995b) 'Alcohol in Hebridean culture: 16th–20th century,' *Addiction* 90: 279–90.

Dean, A. (in press) 'History, culture and substance use in a rural Scottish community,' *Substance Use and Misuse* Special Edition.

de Arcangelis, L. (1987) 'Fractal dimensions in three dimensional Kauffman cellular automata,' *Journal of Physics A: Letters* 20: 369.

Derrida, B. and Pomeau, Y. (1986) 'Random networks of automata: a simple annealed approximation,' *Europhysics Letters* 1: 45.

Derrida, B. and Stauffer, D. (1986) 'Phase transitions in two-dimensional Kauffman cellular automata,' *Europhysics Letters* 2: 739.

Devor, E. J., Cloninger, C. R., Hoffman, P. L. and Tabakoff, B. (1991) 'A genetic study of platelet adenylyl cyclase activity: evidence for a single major locus effect in fluoride stimulated activity,' *American Journal of Human Genetics* 49.2: 372–7.

Durkheim, E. (1933) *The Division of Labour in Society*. New York: Macmillan.

Eckmann, J. P. and Ruelle, D. (1985) 'Ergodic theory of chaos and strange attractors,' *Reviews of Modern Physics* 57.3: 617–56.

Edelman, G. M. (1989) *The Remembered Present: A Biological Theory of Consciousness*. New York: Basic Books.

Edelman, G. (1992) *Bright Air, Brilliant Fire: On the Matter of the Mind*. London: Penguin.

Edwards, G. and Gross, M. M. (1976) 'Alcohol dependence: provisional description of a clinical syndrome,' *British Medical Journal* 6017.1: 1058–61.

Ehlers, C. L. (1992) 'The new physics of chaos: can it help us understand the effects of alcohol?' *Neuroscience* 16.4: 267–272.

Fulker, D. W. and Eysenck, H. J. (1979) 'Nature and nurture: heredity,' in H. J. Eysenck (ed.) *The Structure and Measurement of Intelligence*. New York: Springer.

Gleick, J. (1988) *Chaos. Making a New Science*. London: Cardinal.

Goodwin, D. W. (1989) 'Biological factors in alcohol use and abuse: implications for recognizing and preventing alcohol problems in adolescence,' *International Review of Psychiatry* 1: 41–9.

Goodwin, D. W., Schulsinger, F., Hermansen, L., Guze, S. B. and Winokur, G. (1973) 'Alcohol problems in adoptees raised apart from alcoholic biological parents,' *Archives of General Psychiatry* 28: 238–43.

Goodwin, D. W., Schulsinger, F., Moller, N., Hermansen, L., Winokur, G. and Guze, S. B. (1974) 'Drinking problems in adopted and nonadopted sons of alcoholics,' *Archives of General Psychiatry* 31: 164–9.

Goodwin, D. W., Schulsinger, F., Knop, J., Mednick, S. and Guze, S. B. (1977) 'Alcoholism and depression in adopted-out daughters of alcoholics,' *Archives of General Psychiatry* 34: 751 5.

Gora-Maslak, G., McClearn, G. E., Crabbe, J. C., Phillips, T. L., Belknap, J. K. and Plomin, R. (1991) 'Use of recombinant inbred strains to identify quantitative trait loci in psychopharmacology,' *Psychopharmacology* 104: 413–24.

Gregersen, H. and Sailer, L. (1993) 'Chaos theory and its implications for social science research,' *Human Relations* 46.7: 777–802.

Grisel, J. E. and Crabbe, J. C. (1995) 'Quantitative trait mapping,' *Alcohol Health and Research World* 19.3: 220–7.

Grunberg, N. E. (1994) 'Overview: biological processes relevant to drugs of dependence,' *Addiction* 89: 1443–6.

Habermas, J. (1984) *The Theory of Communicative Action*. London: Heinemann.

Heath, A. C. (1995) 'Genetic influences on alcoholism risk,' *Alcohol Health and Research World* 19.3: 166–71.

Hiller-Sturmhoefel, S., Bowers, B. J. and Wehner, J. M. (1995) 'Genetic engineering in animal models,' *Alcohol Health and Research World* 19.3: 206–13.

Hobbs, J. (1991) 'Chaos and indeterminism,' *Canadian Journal of Philosophy* 21.2: 141–64.

Holman, R. B. (1994) 'Biological effects of central nervous system stimulants,' *Addiction* 89.11: 1435–42.

Holmes, T. H. (1979) 'Developments and application of a quantitative measure of life-change magnitude,' in J. E. Barrett (ed.) *Stress and Mental Disorder*. New York: Raven.

Hrubec, Z. and Omenn, G. S. (1981) 'Evidence of genetic predisposition to alcohol cirrhosis and psychosis: twin concordances for alcoholism and its biological points by zygosity among male veterans,' *Alcoholism: Clinical and Experimental Research* 5: 207–15.

Jaffe, J. H. (1989) 'Addictions: what does biology have to tell?' *International Review of Psychiatry* 1: 51–61.

Jellinek, E. M. (1960) *The Disease Concept of Alcoholism*. New Haven, CT: Hillhouse Press.

Kaij, L. (1960) *Alcoholism in Twins: Studies on the Etiology and Sequels of Abuse of Alcohol*. Stockholm: Almqvist and Wiksell International.

Kauffman, S. A. (1993) *The Origins of Order. Self-organization and Selection in Evolution*. Oxford: Oxford University Press.

Koskenvuo, M., Langinvaino, J., Kaprio, J., Lonnqvist, J. and Tienari, P. (1984) 'Psychiatric hospitalization in twins,' *Acta Geneticae Medicae et Gemellologiae* 33: 321–32.

Lader, M. (1994) 'Biological processes in benzodiazepine dependence,' *Addiction* 89.11: 1413–18.

Lemaire, A. (1977) *Jacques Lacan*. London: Routledge and Kegan Paul.

Lewis, C. S. (1990) *That Hideous Strength. The Cosmic Trilogy*. London: Pan Books.

Lichtenstein, P. and Pedersen, N. L. (1991) 'Genetic analyses of socioeconomic status in twins reared apart and twins reared together,' *Behavior Genetics* 21: 728.

Lichtenstein, P., Pedersen, N. L. and McLearn, G. E. (1992) 'Genetic and environmental predictors of socioeconomic status,' *Behavior Genetics* 22: 731–2.

Littleton, J. and Little, H. (1994) 'Current concepts of ethanol dependence,' *Addiction* 89.11: 1397–412.

Loehlin, J. C. (1992) *Latent Variable Models: An Introduction to Factor, Path and Structural Analysis*, 2nd edn. Hillsdale, NJ: Lawrence Erlbaum.

Lorenz, E. N. (1963) 'Deterministic nonperiodic flows,' *Journal of Atmospheric Science* 29: 130–41.

Loye, D. (1995) 'Prediction in chaotic social, economic, and political conditions: the conflict between traditional chaos theory and the psychology of prediction, and some implications for general evolution theory,' *World Futures* 44: 15–31.

McBurnett, M. (1996) 'Probing the underlying structure in dynamical systems: an introduction to spectral analysis,' in E. Elliott and L. D. Kiel (eds) *Chaos Theory in the Social Sciences: Foundations and Applications*. Ann Arbor: University of Michigan Press.

McGregor, D. (1938) 'The major determinants of the prediction of social events,' *Journal of Abnormal and Social Psychology* 33: 179–204.

Manktelow, K. I. and Over, D. E. (1987) 'Reasoning and rationality,' *Mind and Language* 2: 199–219.

May, R. M. (1976) 'Simple mathematical models with very complicated dynamics,' *Nature* 261: 459–67.

Mendel, G. (1865) 'Versuche uber pflanzen-hybriden,' *Bruenn Natural History Society*, Vol. 4: 3–47.

Mills, C. W. (1970) *The Sociological Imagination*. Harmondsworth: Penguin.

Murray, R. M., Clifford, C. and Gurlin, H. M. (1983) 'Twin and alcoholism studies,' in M. Gallanter (ed.) *Recent Developments in Alcoholism*, Vol. 1. New York: Gardiner Press.

Neale, M. C. and Cardon, L. R. (1992) *Methodology for Genetic Studies of Twins and Families*. Dordrecht: Kluwer Academic.

Neiderheiser, J. M., Plomin, R., Lichtenstein, P., Pedersen, N. L. and McClearn, G. E. (1992) 'The influence of life events on depressive symptoms over time,' *Behavior Genetics* 22: 740.

Orford, J. (1985) *Excessive Appetites: A Psychological View of Addictions*. Chichester: Wiley.

Partanen, J. K., Brunn, K. and Markkanen, T. (1966) *Inheritance of drinking behaviour. A study on intelligence, personality and use of alcohol in adult twins*. Helsinki: Finnish Foundation for Alcohol Studies.

Plomin, R. (1990) 'The role of inheritance in behavior,' *Science* 248: 183–8.

Plomin, R. (1993) 'Nature and nurture: perspective and prospective,' in R. Plomin and G. E. McClearn (eds) *Nature, Nurture and Psychology*. Washington, DC: American Psychological Association.

Plomin, R. (1994) *Genetics and Experience: The Interplay between Nature and Nurture*. London: Sage.

Plomin, R., De Fries, J. C. and Loehlin, J. C. (1977) 'Genotype–environment interaction and correlation in the analysis of human behavior,' *Psychological Bulletin* 84: 302–22.

Plomin, R., McClearn, G. E., Pedersen, N. L., Nesselroade, J. R. and Bergman, C. S. (1988) 'Genetic influence on childhood family environment perceived retrospectively from the last half of the life span,' *Developmental Psychology* 24: 738–45.

Plomin, R., McClearn, G. E., Pedersen, N. L., Nesselroade, J. R. and Bergeman, C. S. (1989) 'Genetic influence on adults' ratings of their current family environment,' *Journal of Marriage and the Family* 51: 160–3.

Plomin, R., McClearn, G. E., Gora-Maslak, G. and Neiderheiser, J. M. (1991) 'Use of recombinant strains to detect quantitative trait loci associated with behaviour,' *Behaviour Genetics* 21: 99–116.

Plotkin, H. (1994) *The Nature of Knowledge*. London: Penguin.

Reiss, D., Plomin, R. and Hetherington, E. M. (1991) 'Genetics and psychiatry: an unheralded window on the environment,' *American Journal of Psychiatry* 148.3: 283–91.

Rodriguez, L. A., Plomkin, R., Blizard, D. A., Jones, B. C. and McClearn, G. E. (1995) 'Alcohol acceptance, preference and sensitivity in mice II: quantitative trait loci mapping analysis using BXD recombinant inbred strains,' *Alcoholism: Clinical and Experimental Research* 19.2: 367–73.

Roe, A. (1944) 'The adult adjustment of children of alcoholic parents raised in foster homes,' *Journal of Studies on Alcohol* 5: 378–93.

Romanov, K., Kaprio, J. and Rose, R. J. (1991) 'Genetics of alcoholism: effects of migration on concordance rates among male twins,' *Alcohol and Alcoholism* 1. Suppl.: 137–40.

Ruelle, D. (1980) 'Les attracteurs étranges,' *La Recherche* 108: 132.

Ruelle, D. (1983) 'Five turbulent problems,' *Physica* 7: 40–4.

Ruelle, D. (1989) *Chaotic Evolution and Strange Attractors.* New York: Cambridge University Press.

Ruelle, D. (1993) *Chance and Chaos.* London: Penguin.

Saperstein, A. M. (1991) 'The long peace – results of a bi-polar competitive world?' *Journal of Conflict Resolution* 35: 68–79.

Saperstein, A. M. (1996) 'The prediction of unpredictability: applications of the new paradigm of chaos in dynamical systems to the old problem of the stability of a system of hostile nations,' in E. Elliott and L. D. Kiel (eds) *Chaos Theory in the Social Sciences. Foundations and Applications.* Ann Arbor: University of Michigan Press.

Schatzki, T. R. (1993) 'Wittgenstein: mind, body and society,' *Journal for the Theory of Social Behavior* 23.3: 285–313.

Searle, J. R. (1995) *The Rediscovery of the Mind.* Massachusetts: MIT Press.

Shaw, R. S. (1981) 'Strange attractors, chaotic behavior, and information flow,' *Zeitschrift für Natürforschung* 36a: 80.

Sher, K. J., Bylund, D. B., Walizer, K. S., Hartmann, J. and Ray-Prenger, C. (1994) 'Platelet monoamine oxidase (MAO) activity: personality, substance use and the stress response dampening effect of alcohol,' *Experimental and Clinical Psychopharmacology* 2.1: 53–81.

Stewart, I. (1990) *Does God Play Dice? The Mathematics of Chaos.* London: Penguin.

Tabakoff, B., Hoffman, P. L., Lee, J. M., Saito, T., Willard, B. and DeLeon-Jones, F. (1988) 'Differences in platelet enzyme activity between alcoholics and non-alcoholics,' *New England Journal of Medicine* 31.3: 134–9.

Tambs, K., Sundat, J. M., Magnus, P., Magnus, K. B. and Berg, F. (1989) 'Genetic and environmental contributions to the covariance between occupational status, educational attainment and IQ: a study of twins,' *Behavioral Genetics* 19: 209–22.

Taubman, P. (1976) 'The determinants of earnings: genetics, family and other environment. A study of white male twins,' *American Economic Review* 66: 858–70.

Teasdale, T. W. (1979) 'Social class correlations among adoptees and their biological and adoptive parents,' *Behavior Genetics* 9: 103–14.

Teasdale, T. W. and Owen, D. R. (1981) 'Social class correlations among separately adopted siblings and unrelated individuals adopted together,' *Behavioral Genetics* 11: 577–688.

Tsuang, M. T., Lyons, M. J., Elsen, S. A., True, W. T., Goldberg, J. and Hendersen, W. (1992) 'A twin study of drug exposure and initiation of use,' *Behavior Genetics* 22: 756.

Valliant, G. E. (1983) *The Natural History of Alcoholism.* Cambridge, MA: Harvard University Press.

Waltman, C., Levine, M. A., McCaul, M. E., Svikis, D. S. and Wand, G. S. (1993) 'Enhanced expression of the inhibitory protein Gi2a and decreased activity of adenylyl cyclase in lymphocytes of abstinent alcoholics,' *Alcoholism: Clinical and Experimental Research* 17.2: 315–20.

Wason, P. C. (1972) 'Reasoning,' in B. M. Foss (ed.) *New Horizons in Psychology 1*. Harmondsworth: Penguin.

Weisbuch, G. and Stauffer, D. (1987) 'Phase transition in cellular random Boolean nets,' *Journal Physique* 48: 11.

Wise, R. A. (1988) 'The neurobiology of craving: implications for the understanding and treatment of addiction,' *Journal of Abnormal Psychology* 2: 118–32.

Wittgenstein, L. (1973) 'The limits of my language mean the limits of my world,' in M. Douglas (ed.) *Rules and Meanings*. Harmondsworth: Penguin.

Wolf, A. (1986) 'Quantifying chaos with Lyapunov exponents,' in A. Holden (ed.) *Chaos*. Princeton, NJ: Princeton University Press.

Index

theoretical
formulation
↓
empirical